Glenn's Gas Guzzler's Guide

by Harold T. Glenn

Glenn's Gas Guzzler's Guide

by Harold T. Glenn

Henry Regnery Company
Chicago

Library of Congress Cataloging in Publication Data

Glenn, Harold T.
    Glenn's gas guzzler's guide

    I. Automobiles—Fuel consumption. I. Title.
II. Title: Gas guzzler's guide.
TL154.G57   629.2'53   75-6979
ISBN 0-8092-8192-9

Copyright © 1975 by Harold T. Glenn
All rights reserved. This volume
may not be reproduced in whole or
in part in any form without
written permission from the publisher.

Published by Henry Regnery Company
180 North Michigan Avenue, Chicago, Illinois 60601

Printed in the United States of America
Library of Congress Catalog Card Number: 75-6979
International Standard Book Number: 0-8092-8192-9

foreword

This book is written for the "average" car owner (Mr. or Ms.) who wants to save money in these days of inflated prices for gasoline and automobile repairs. Most of the book is devoted to fuel-saving tips, basic tests, and simple maintenance adjustments using only a screwdriver and a pair of pliers. A simple vacuum gauge (about $10.00) is enough to handle all of the engine tests and tune-up adjustments. Elaborate test equipment and mechanical expertise are not needed.

acknowledgments

The author wishes to express his appreciation to the many automobile manufacturers who supplied him with technical information and illustrations.

Special thanks are due to my wife, Anna Glenn, for her gracious and devoted assistance in helping to proofread the manuscript and galley proofs and for her many constructive criticisms. Appreciation is expressed for Mark Tsunawaki's contribution to the artwork of this book.

Harold T. Glenn

Books by HAROLD T. GLENN

Youth at the Wheel
Safe Living
Automechanics
Glenn's Auto Troubleshooting Guide
Exploring Power Mechanics
Automobile Engine Rebuilding and Maintenance
Automobile Power Accessories
Glenn's Auto Repair Manual
Automotive Smog Control Manual
Glenn's Emission-Control Systems
Glenn's Tune-Up and Repair Manual for American and Imported Car Emission-Control Systems
Glenn's Foreign Car Repair Manual
Glenn's Triumph Repair and Tune-Up Guide
Glenn's Alfa Romeo Repair and Tune-Up Guide
Glenn's Austin, Austin-Healey Repair and Tune-Up Guide
Glenn's Sunbeam-Hillman Repair and Tune-Up Guide
Glenn's MG, Morris and Magnette Repair and Tune-Up Guide
Glenn's Volkswagen Repair and Tune-Up Guide
Glenn's Volkswagen Repair and Tune-Up Guide (Spanish Edition)
Glenn's Mercedes-Benz Repair and Tune-Up Guide
Glenn's Foreign Carburetors and Electrical Systems Guide
Glenn's Renault Repair and Tune-Up Guide
Glenn's Jaguar Repair and Tune-Up Guide
Glenn's Volvo Repair and Tune-Up Guide
Glenn's Peugeot Repair and Tune-Up Guide
Glenn's Fiat Repair and Tune-Up Guide
Glenn's Toyota Tune-Up and Repair Guide
Glenn's Mazda Tune-Up and Repair Guide
Glenn's Chrysler Outboard Motor Repair and Tune-Up Guide for 1 & 2 Cylinder Engines
Glenn's Chrysler Outboard Motor Repair and Tune-Up Guide for 3 & 4 Cylinder Engines
Glenn's Evinrude Outboard Motor Repair and Tune-Up Guide for 1 & 2 Cylinder Engines
Glenn's Evinrude Outboard Motor Repair and Tune-Up Guide for 3 & 4 Cylinder Engines
Glenn's Johnson Outboard Motor Repair and Tune-Up Guide for 1 & 2 Cylinder Engines
Glenn's Johnson Outboard Motor Repair and Tune-Up Guide for 3 & 4 Cylinder Engines
Glenn's McCulloch Outboard Motor Repair and Tune-Up Guide
Glenn's Mercury Outboard Motor Repair and Tune-Up Guide
Glenn's Sears Outboard Motor Repair and Tune-Up Guide
Honda One-Cylinder Repair and Tune-Up Guide
Glenn's Honda Two-Cylinder Repair and Tune-Up Guide
Suzuki One-Cylinder Tune-Up and Repair Guide
Yamaha Enduro Tune-Up and Repair Guide
Triumph Two-Cylinder Motorcycle Tune-Up and Repair Guide
Glenn's Chevrolet Tune-Up and Repair Guide
Glenn's Chevrolet Camaro Tune-Up and Repair Guide
Glenn's Chevrolet Vega Tune-Up and Repair Guide
Glenn's Ford/Lincoln/Mercury Tune-Up and Repair Guide
Glenn's Chrysler/Plymouth/Dodge Tune-Up and Repair Guide
Glenn's Pontiac Tune-Up and Repair Guide
Glenn's Pontiac Firebird Tune-Up and Repair Guide
Glenn's Oldsmobile Tune-Up and Repair Guide
Glenn's Buick Tune-Up and Repair Guide
Glenn's Complete Bicycle Manual

contents

1 ARE YOU GETTING 164.5 MILES PER GALLON? 1
  Average fuel consumption costs 3
  164.5 miles per gallon 4
  How much are you getting? 6
  The basic mileage test 7

2 ENGINE REPAIRS FOR ECONOMY 10
  Tuning the enine 13
  Mechanical conditions 15
      Heat-control valve 15
      Valve adjustment 16
  Ignition system 16
      Spark plugs 17
          Wasting fuel 20
          Replacing spark plugs 22
          Selecting the right spark plugs 22
          Removing the spark plugs 23
      Testing the compression 25
      Servicing the distributor 26
          Dwell or cam angle 32
          Automatic advance mechanisms 33
          Vacuum advance 36
          Thermal Vacuum-Switching valve 39
          Servicing the automatic advance
              mechanisms 40

Replacing the distributor 41
Setting the ignition timing 43
Power-timing the engine 45
Fuel system service 49
    Air filters 52
    Thermostatically-controlled air cleaner 57
        Service procedures 59
Gas mileage improvers 59
Carburetors 62
    Service procedures 64
    Carburetor kits 65
    Carburetor circuits 69
        Float system 69
        Idle system 71
        Main metering system 72
        Power system 72
        Accelerating system 73
        Automatic choke 74
    Carburetor adjustments 77
        Emission-controlled engines 79
        Lean-drop method 79
Emission-control systems 81
    Crankcase emission-control system 82
        System quick test 82
        Service procedures 84
Exhaust emission-control systems 86
    Transmission-Controlled Spark (TCS) 86
        System basic tests 89
        System isolation test 93
    Spark-Delay systems (SDS) 97
        System basic test 98
        Service procedures 99
    Exhaust-Gas Recirculation (EGR) 100
        System quick tests 101

3 VEHICLE MAINTENANCE FOR ECONOMY 103
  Vehicle weight 104
    Extra weight 105

Wind resistance   106
    Roof and bicycle racks   108
    Vinyl top   108
Rolling resistance   108
    Tires   109
    Reducing tire rolling resistance   114
    Wheel alignment   117
    Disc brakes   117
Buying gasoline wisely   118
    Octane numbers   121
Buying oil wisely   122

4 HOW TO DRIVE FOR ECONOMY   123
Improving your driving skills   125
    Starting a cold engine   126
        Thermostat   128
    Preplanning   130
    Excessive idling   130
    Getting underway   131
    Pacing traffic lights   132
    Be an "Old Smoothie"   133
    Driving the Freeways   135
    Driving in hilly country   136
Accessories that affect gas mileage   139
    Vacuum gauge   139
    Wooden block cruise control   142
    Cruise control   145
    Air conditioner   147

APPENDIX   149
    EPA labeling program   149
        The specific label   150
        The comparative label   151
    1975 gas mileage guide for new car buyers   151
    Spark plug torque specifications   157
    Gas mileage improvement schedules   158

INDEX   162

CHAPTER 1
Are you getting 164.5 miles per gallon?

America's love affair with the automobile is over! Hot-rodding, jack-rabbit starts, and tire-screeching stops are frowned upon as a national conservation measure and as a personal economic necessity. Let's face it, the days of cheap energy are gone, never to return, and the automobile, while still a basic necessity, will become more utilitarian.

With Washington legislating the manufacture of cars, the Arabs controlling the flow of fuel, and the EPA (Environmental Protection Agency) regulating the automobile's use and the quality of the engine's exhaust, the modern automobile has become a complex mechanism that is fast becoming neither economical nor legally possible of modification for increasing the engine's power and efficiency.

With the Arabs using fuel as a power-producing weapon, turning it on and off as the political situation changes, and the dependence of the rest of the World on petroleum products, we find that these small Mid-eastern nations are doing as they wish with impunity. And now we have the U.S. government entering the energy picture, wanting us to become independent of the awesome power of the Arabs.

Political decisions made in Washington can only lead to more stringent restrictions in the use of fuel by private citizens. Whether it comes in the form of rationing or higher taxes on gasoline to restrict car use, it means that the individual driver must consider his position a very tenuous one. You are the one who is going to pay the price of the power struggle between our government and the Arabs.

To this end, we are exhorted to be patriotic and to save all of the fuel possible by using public transportation and car-pooling to release us from dependence on the Arabs and, for you and me, it is

Distributed by *Los Angeles Times* SYNDICATE

Read this book first, you may be able to save it.

FUEL CONSUMPTION COSTS 3

essential to become more economy-minded to save all of the "bucks" possible in operating our own cars in today's inflationary atmosphere. Both of these objectives are one and the same, and that is the purpose of this book; to save you money in operating your car and to avoid wasting that precious comodity, gasoline.

This book is a practical how-to-do-it guide for lowering your fuel consumption and, thereby, reducing your costs for operating your automobile. We will not take up space with fixed-cost items, such as buying a new car with power-saving features, depreciation, insurance, and taxes, but we will devote this book entirely to gas-consumption variables on your present car, **those which you can control**: the engine, the vehicle, and your driving habits.

Most drivers run their vehicles about 12,000 miles per year and, at an average gas consumption rate of 12 mpg, are burning about 1,000 gallons of gasoline in that time. With fuel costs at about 60¢ per gallon, that's almost $600.00 per year for energy. If fuel prices rise to $1.00 per gallon (don't bet it won't; its well over that in most European countries right now), then your fuel costs will be at least $1,000.00 per year. If you can realize a savings of 10% by following some of the tuning or driving hints in this book, you'll be saving $100.00 per year. A 15% savings will be putting $150.00 back into your pocket. Savings of 25% should not be unusual, and this means that you'll get back 1/4 of your fuel costs, or $250.00 per year.

AVERAGE FUEL CONSUMPTION AND COSTS

MILES DRIVEN	FUEL CONSUMED @ 12 MPG (Gallons)	FUEL COSTS @ 60¢ PER GALLON (Dollars)
Yearly:		
12,000	1,000	600.00
Monthly:		
1,000	83.3	50.00
Daily:		
33.3	2.78	1.67

This table is based on a yearly average of 12,000 miles, with the vehicle obtaining an average of 12 mpg and fuel costing 60¢ per gallon.

4 GAS GUZZLER'S GUIDE

164.5 MILES PER GALLON

The maximum fuel potential of a vehicle, at a constant road speed of 20 mph, is 164.5 mpg (miles per gallon), but this is a mathematical analysis which works only in theory. Why, then, do we discuss it? If you visualize the concepts about this theoretical maximum, then you will be able to see where some of your precious gas is going and why our item-by-item chipping away at the losses will result in significant gains.

Bear with us. One BTU = 778.26 foot-pounds of work. One horsepower = 33,000 foot-pounds of work per minute. Therefore, 2,540 BTU = 1 hp for 1 hour. Gasoline contains the energy equivalent of 115,000 BTU per gallon. So. . . 0.0221 gallons of fuel will produce 1 hp **if all the energy is converted into work.** It

The Federal Energy Administration's concept of the average fuel-guzzler's automobile. How does yours compare?

ARE YOU GETTING ENOUGH?   5

requires 5.5 brake horsepower to maintain a constant speed of 20 mph. The theoretical maximum fuel economy becomes 20 divided by 0.12155 (5.5 × 0.0221), or 164.5 mpg.

But this is only theory. No engine operates at 100% efficiency. So now we start to "take it off." In converting gasoline energy into heat energy and then into mechanical energy, about 45% of the total is lost in the burned gases which are exhausted into the atmosphere in order to make room for the next fuel/air charge and through radiation from hot engine parts. 20% heat energy is dissipated to the coolant. Another 5% is lost through friction between rotating and reciprocating engine parts. So your engine, at most efficient operation at full throttle, is really only about 30% efficient, which calculates out to 49.35 mpg (164.5 × 30%), which is still not bad. **But are you getting it?** Read on.

"Well, there goes the neighborhood . . ."

HOW MUCH ARE YOU GETTING?

"Gas mileage", as one sage put it, "is a subject about which there is a great deal of loose talk, but little concrete knowledge." Everyone knows that his car is getting "X" miles per gallon but, in reality, he does not actually know unless his testing program is scientifically carried out, as very few of them are. In general, most drivers will record the mileage between fill-ups to determine the

Good to the last drop!

mpg (miles per gallon) by dividing the miles driven by the gallons required to fill the tank, **but this is not accurate!**

There are just too many variables for a driver to say with any degree of certainty that he is getting "X" mpg. If you keep track of each fill-up and compute the mpg, you will find that it varies each time, and sometimes by a significant amount. Why should there be such descrepancies? They occur because of the many variables: (1) Is the tank filled to exactly the same level each time? (2) Is the vehicle absolutely level at each fill-up? (3) Did you drive exactly the same route during each test? (4) Were the traffic conditions exactly the same? (5) Did you stop at exactly the same number of traffic lights? (6) Did you make the same number of cold starts during each test, where a cold engine and the action of the automatic choke make the engine operate least efficiently as far as fuel consumption is concerned?

How, then, can you determine your actual mileage? How do you determine whether the lowered mileage you are presently experiencing is caused by the engine, vehicle, or your driving habits? It's easy to make some tests to isolate the trouble so you can make corrections, but you must make some scientifically accurate tests first.

THE BASIC MILEAGE TEST

To determine your car's actual mileage potential, without changing anything at this time to obtain a base upon which you can compare improvements accurately, make the first mileage test precisely as follows: First make sure that your tires are inflated to the recommended air pressure. Fill the gas tank to the top, as full as possible. See to it yourself that every drop that can be squeezed into the tank is there. **CAUTION: Some late-model cars with evaporative emission-control systems have small tanks inside the main gas tank with bleed holes in them to trap air during the filling process; more fuel can be added if you wait a short time for the secondary tank to fill partially.** Pay attention to the pump from which you get the gas and your exact position at the pumping island so that you can return to the same place to fill the tank at the end of the test.

Now, drive the vehicle along a test route that covers at least 100 miles and do it as close to 35 mph as possible. **The longer the test route, the more accurate the results will be.** Select a route with a minimum of lights, stop signs, and traffic so that you can repeat the test after having made some of the suggested changes. In this way, you can make accurate comparisons and see if you have made a significant mileage improvement. This basic test must be made by varying the throttle as little as possible so that you are testing **the mileage-potential of the vehicle**, leaving out as many variables as possible.

At the end of the test, return to the same filling station. Fill the

Make sure that the car is level and at exactly the same position at the gas pump.

gas tank at the same pump and to the same level as when you started the test. Also, wait the same amount of time to allow the secondary tank inside of the main one to fill. Now, calculate your gas mileage by dividing your gallons needed to fill the tank into the mileage driven. Calculate it accurately, using the tenth-of-a-mile figure on the odometer and the tenth-of-a-gallon reading on the gas pump, because we are going to establish **the base mileage figure that the car can deliver**.

Even this figure will vary if you run the same test again, because you never can control the number of times you had to slow for a pedestrian or to stop for an unexpected traffic light. However, if you do try to control the variables, you will be better able to determine just how much mileage the car can possibly deliver at a relatively constant speed. This figure may surprise you, because it will most certainly be higher than you have ever gotten. But remember that you are testing the vehicle's mileage potential to be used as a base to compare the results that suggested changes will make. Careful attention to accuracy is essential at this point.

Now, let's start taking it off!

The mileage you get is greatly influenced by the type of driver you are and the traffic conditions under which you drive. Remember that you are in the driver's seat and that a heavy foot wastes gasoline.

CHAPTER 2
engine repairs for economy

Your engine consumes fuel to propel your car, and it must do this as efficiently as possible to realize the best economy. All parts of an engine are subject to wear as mileage accumulates, and this is why you should have periodic tune-ups. A tune-up consists of replacing worn parts and making adjustments to the ignition and fuel systems to restore the pep, power, and gas mileage.

The combustion of gasoline and air is used to power all engines. At the left we see a steam engine (external combustion) where the flames are under the boiler. At the right, we see an internal-combustion engine where the flame is used to push a piston down.

FOUR-STROKE CYCLE ENGINE 11

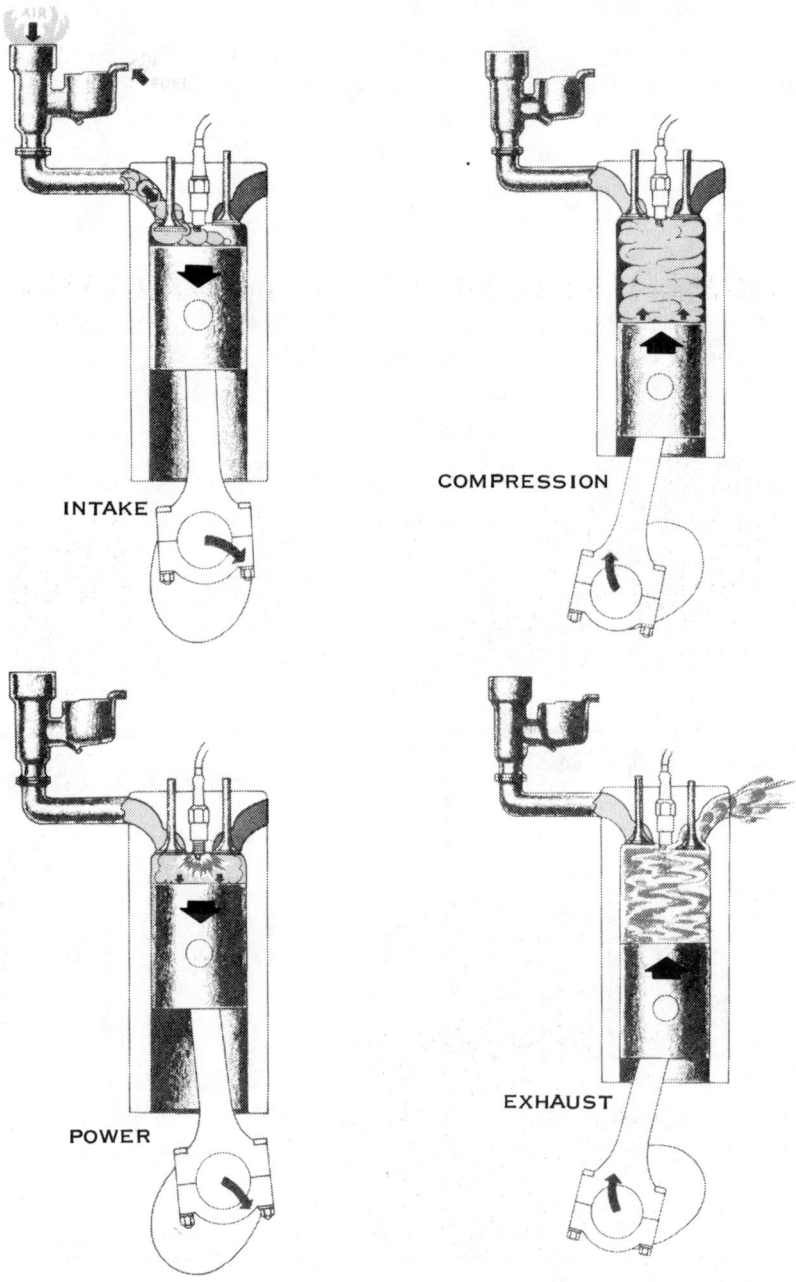

The four strokes of a four-cycle internal-combustion engine. Note the sequence of events: (1) Intake stroke; air and gasoline are mixed and drawn into the combustion chamber past the opened intake valve as the piston descends. (2) Compression stroke; both valves are closed as the piston moves up to trap and compress the flamable mixture. (3) Power stroke; the spark plug ignites the mixture and drives the piston down to propel the vehicle. Both valves are closed during this stroke. (4) Exhaust stroke; the piston rises and forces the burned gas out past the opened exhaust valve.

There's no way of telling just how much fuel you're wasting by postponing a needed tune-up. Because the longer you go, the worse the engine runs, and the greater your fuel loss. If you notice that the engine is running rougher, not starting promptly, and not getting the mileage it once did, then you probably need a tune-up. (You can confirm this by rerunning the basic test outlined in the first chapter of this book.)

In any event, it is money in your pocket if you do your own work and perform a tune-up every 5,000 miles. If you have to have the work done, you must make a trade-off in considering just how much it is going to cost you for the work as compared with your increasingly higher operating costs. Really, it's not too big a deal to do the few things spelled out in this section, **even without elaborate equipment**. All you need are a few basic tools and a vacuum gauge. A timing light is handy, too.

A tune-up may not recover your lost gas mileage if you have some real problems in your fuel or emission-control system; therefore, two sections on these topics follow the tune-up material to check out the rest of the engine's potential fuel-robbing systems.

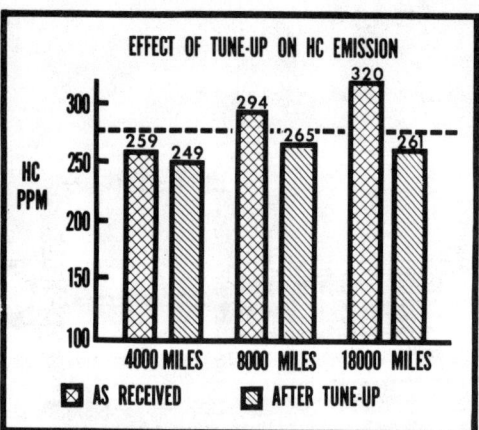 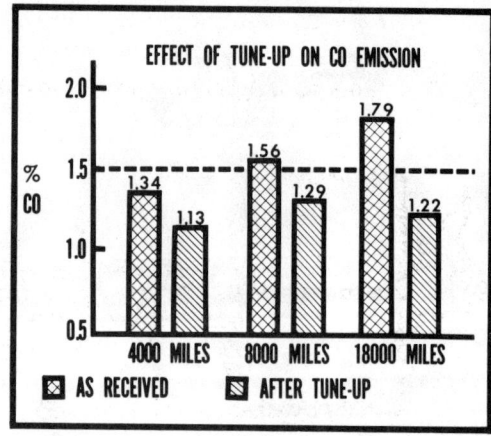

Improved engine performance, better gas mileage, and reduced exhaust emissions result from periodic tune-ups. This is another way of saying that we can clean the air while we save money.

TUNING THE ENGINE   13

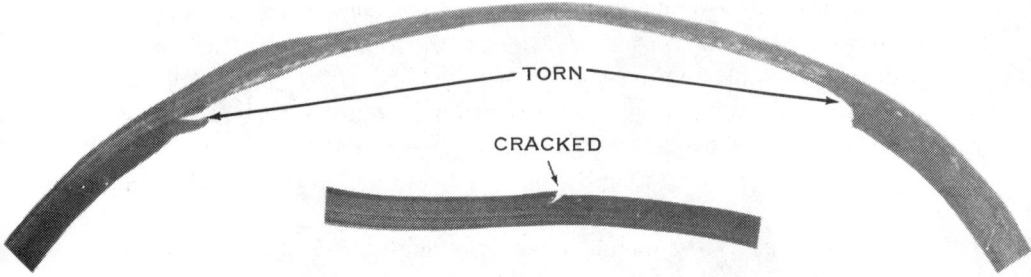

A worn fan belt should be replaced. Check the condition of the belt at each tune-up to avoid breakage, which could cause engine damage. Replace the belt if you can see any cracks in its inner surface.

TUNING THE ENGINE

You must make repairs and adjustments to the ignition and fuel systems and it is considered good trade practice to do these jobs in the order which follows: (1) check the mechanical condition of the engine, (2) make repairs to the ignition and fuel systems, and (3) adjust the carburetor. True, late-model engines, with their emission-control systems, present problems in that you must take into consideration the emission level of the exhaust, therefore, a separate section on tuning emission-controlled engines follows. Use both sections to obtain the most efficient-running engine possible.

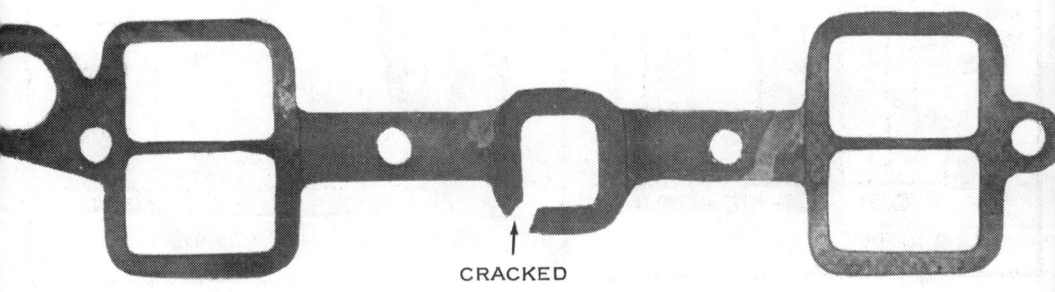

An intake manifold gasket with a section out will lean the air-fuel mixture to two cylinders and cause them to misfire. In addition to upsetting the mixture, it causes the engine to idle roughly. Be sure to tighten all manifold gaskets to avoid air leaks.

14   GAS GUZZLER'S GUIDE

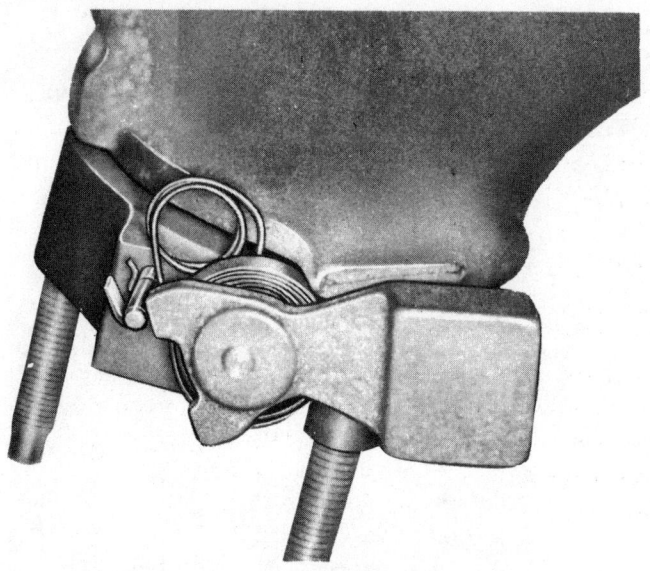

Check the manifold heat control valve to be sure that it is free. A stuck valve can affect engine warm-up and waste gasoline. CAUTION: Always use "Lube Trap" lubricant, never engine oil; otherwise, the valve will stick because of the heat under which it operates.

Uneven tightening of this manifold cracked it. Always use a torque wrench to prevent this from happening.

MECHANICAL CONDITIONS

Before doing any work on the ignition system, it is good practice to tighten the cylinder head nuts and intake manifold bolts to the specified torque. The compression will be tested after removing the spark plugs.

HEAT-CONTROL VALVE

This thermostatically-controlled valve, built into the exhaust manifold, can stick open and result in some non-vaporized fuel entering the combustion chambers. Liquid fuel cannot burn, and this is wasted gasoline. The answer is to use a special "lub trap" lubricant to free up the valve. **CAUTION: Don't use engine oil, or the intense heat of the exhaust will cause it to burn and carbonize.**

This manifold heat control valve was stuck, and the mechanic hammered on the shaft to free it up, which broke the casting.

16 GAS GUZZLER'S GUIDE

VALVE ADJUSTMENT

Some engines have mechanical tappets and some have hydraulic lifters that require an adjustment every tune-up. If the valve lash is not adjusted properly, the valve timing will be off and so will engine performance and gas mileage.

IGNITION SYSTEM

The ignition system consists of a distributor, ignition coil, and spark plugs. Of course, we need a battery to supply the energy for starting and to keep the engine running when the alternator is not charging. Three things must be done to the distributor to restore it

Adjusting the hydraulic valve lifter in a Chevrolet in-line, six-cylinder engine (left) and a V-8 engine. In each case, the rocker arm adjusting nut is turned down until there is zero lash, which can be determined when you can't turn the push rod easily. Then turn the nut down one additional turn to position the lifter in the center of its travel.

SERVICING THE IGNITION SYSTEM 17

to good operating condition. You should replace the contact points, space them properly, and adjust the ignition timing. In addition, you should clean or replace the spark plugs and clean the battery terminals of all corrosion. Generally, the other parts of the ignition system require little attention.

SPARK PLUGS

The spark plug ignites the compressed air-fuel mixture to start combustion. The resulting heat drives the piston down to power the vehicle. The spark plug must do its work under extremely difficult

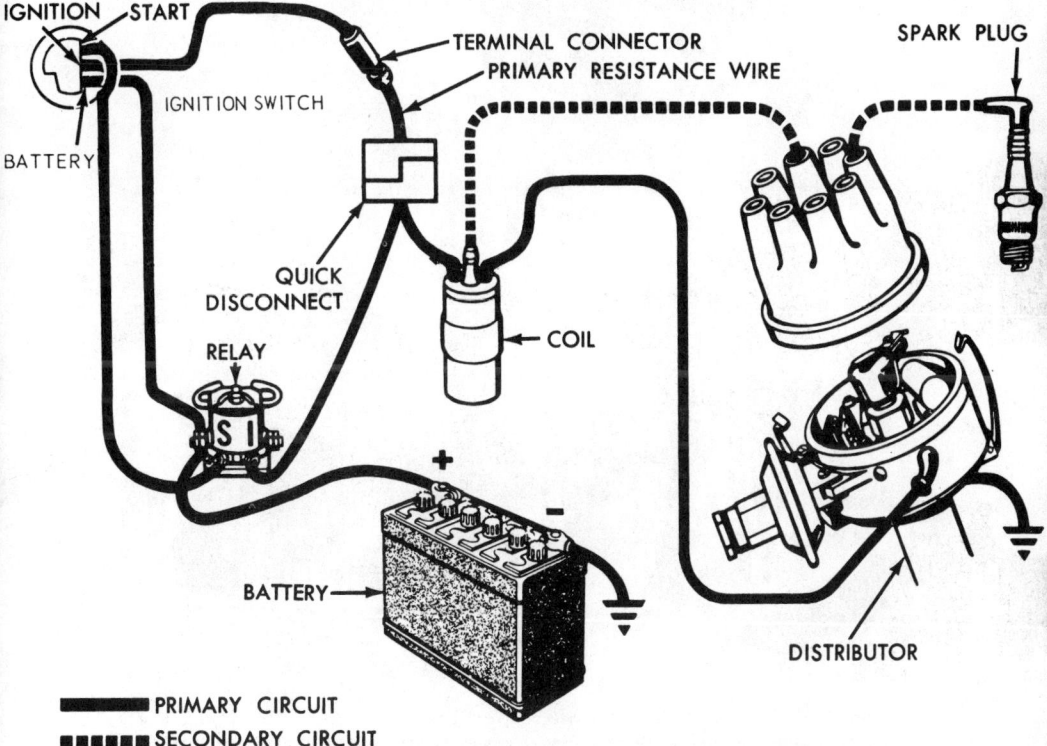

Typical ignition system circuit. Some cars have a ballast resistor in place of the primary resistance wire shown.

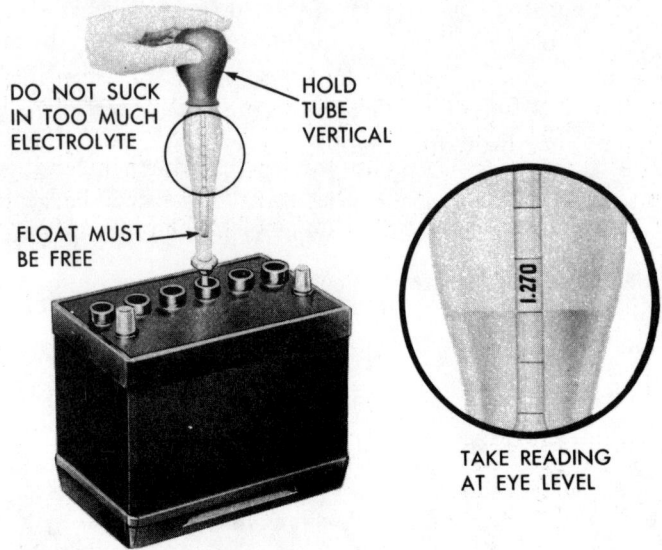

Check the specific gravity of your battery to determine its state of charge. Suck up only enough electrolyte to free the float.

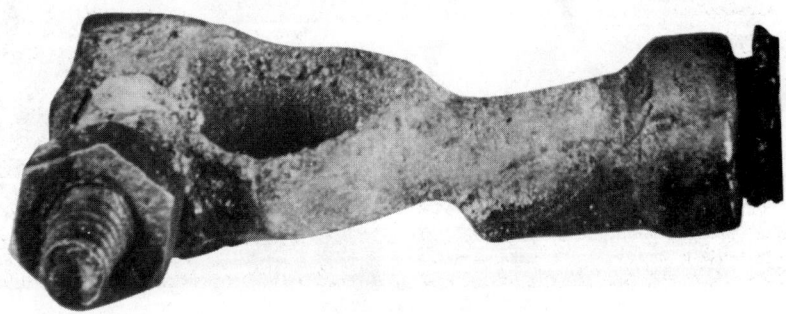

This corroded battery cable connector caused all kinds of electrical problems due to the high voltage developed by the alternator in trying to get across the poor connection. The alternator and regulator were destroyed, the headlamps burned out regularly, and the ignition points were ruined. And the gas mileage suffered, too!

SERVICING THE IGNITION SYSTEM 19

conditions of heat and pressure, delivering a voltage surge across the electrodes high enough to fire the mixture every time.

The combustion temperatures are in the neighborhood of 4,000°F, the spark plug insulator's temperature is about 800°F, and the air-fuel mixture is compressed about 10 times, making the initial working pressures within the combustion chamber about 180 psi. In addition, each spark plug must fire every charge for the engine to develop maximum horsepower for efficient operation and good gas mileage.

With a six-cylinder engine operating at 3,000 rpm, at Freeway driving speed, each spark plug of a six-cylinder engine must ignite a compressed air-fuel charge 250 times per minute. If any of the spark plugs fail, even once, the engine misfires, and that charge of fuel is pumped out of the combustion chamber to the muffler, doing absolutely no work in the process. No wonder, then, we consider continuously firing spark plugs so important in the chain of events that leads to good engine performance and economical operation.

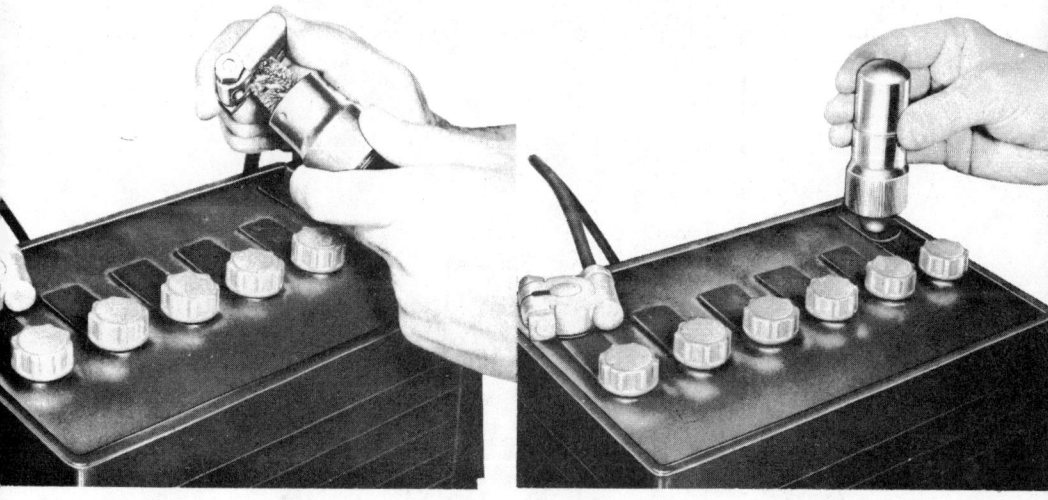

One of the greatest difficulties in the electrical system is the maintenance of the correct operating voltage and this, in turn, is dependent on minimizing resistance between the connections in the charging circuit. One of the most important service procedures is cleaning the battery terminal posts and cable connections of all corrosion. Scrape them until you can see bright metal.

Wasting Fuel

One nonfiring spark plug in an eight-cylinder engine can reduce your gas mileage by 12%, that's wasting $6.00 per month. In two months, you could save enough to replace the entire set, cost-free! In addition, a new set of spark plugs will restore lost engine performance and make engine starting easier. Worth it? You bet!

Unfortunately, you may not be aware that your engine is misfiring, especially if you are driving a vehicle with a V-8 engine. The

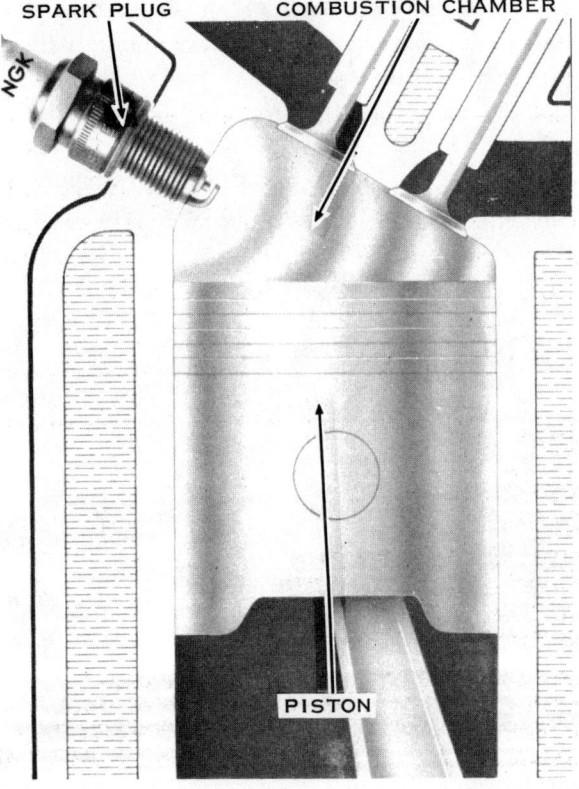

The spark plug is the sole eye-witness to the action which goes on inside of the combustion chamber. Its firing end tells a story of engine-operating conditions.

SERVICING THE SPARK PLUGS 21

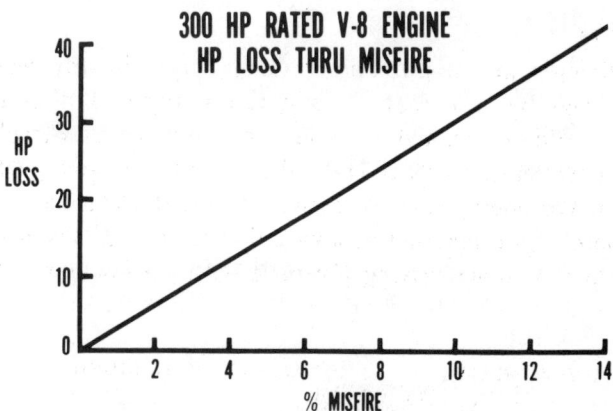

This graph shows the Hp loss as compared with the per cent of engine misfiring.

power flow is so smooth that intermittent misfiring cannot always be detected, unless you lug the engine at very low speeds, where you can feel the jerky operation. Regardless, if there is no spark or there is an intermittent one, then that fuel charge is wasted.

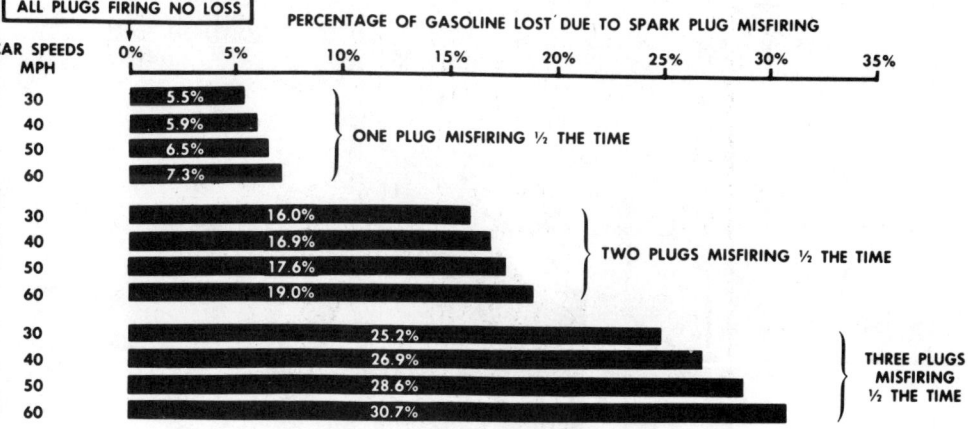

If one spark plug is firing only half of the time, the gas mileage will fall off as much as 7% and this will waste about 6 gallons of gasoline per month. With two spark plugs misfiring, the mileage can drop as much as 20% for a cost of 35¢ per day.

Replacing Spark Plugs

To reduce this waste, you must replace the spark plugs every 10,000 miles (12,000-22,500 miles in 1975 engines using lead-free fuels). Clean them every 5,000 miles to keep them operating efficiently, if you have sand-blast facilities. In the event you don't have the proper equipment, you can recondition spark plugs by removing them, filing the electrode gap square, gapping them to specifications, and replacing them. If the spark plugs are not shorted out or the porcelain insulator cracked, this will restore them to good working order.

Selecting The Right Spark Plugs

Spark plugs are manufactured in a great number of types and heat ranges, which is an exceedingly important factor for the consumer to know about. A hot-type spark plug has a relatively long heat path from the tip of the insulator (which is exposed to combustion) to the coolant. A cold-type, on the other hand, has a short path

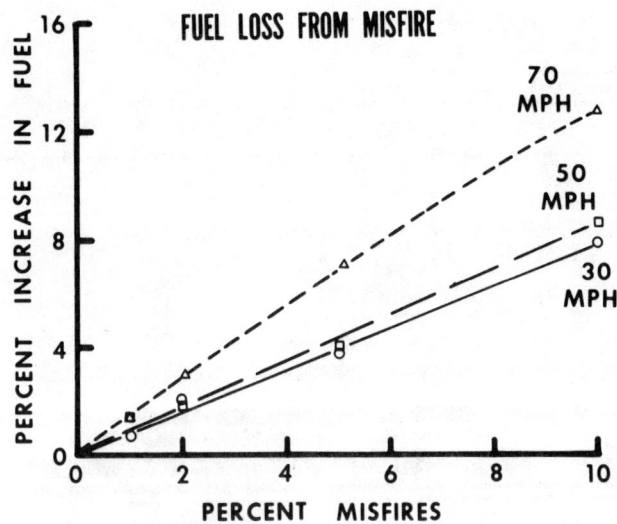

Graph showing the fuel loss caused by spark plug misfire and the percent increase in fuel. a 10% loss in economy costs you about $5.00 per month in wasted gasoline.

SELECTING THE RIGHT SPARK PLUGS

to dissipate this heat energy quickly in order to keep its working conditions within design limits.

When you buy a set of spark plugs, you are depending on a clerk to sell you the proper spark plugs for your engine. In most cases, they will be the correct ones. However, there are times that he might be out of stock and substitute a set that is "equally as good", **in his opinion**. If you are not aware of the proper spark plug designation for your engine, then you can be installing a set with an incorrect heat range rating. How can you tell? Read on.

The heat range specified for your engine is one that fits average running conditions. If you are driving mostly in the city, predominately short trips, then your engine does not warm up enough to reach the designed spark plug insulator temperature, and the plugs can foul and misfire. On the other hand, you may be doing a great deal of high-speed driving, and the temperature of the spark plug insulator could rise above design specifications. This will result in blistering of the insulator, pre-ignition, and rapid electrode wear, again causing engine misfire.

REMOVING THE SPARK PLUGS

Remove the spark plug wires by grasping, twisting, and then pulling on the molded cap only. **CAUTION: Don't pull on the**

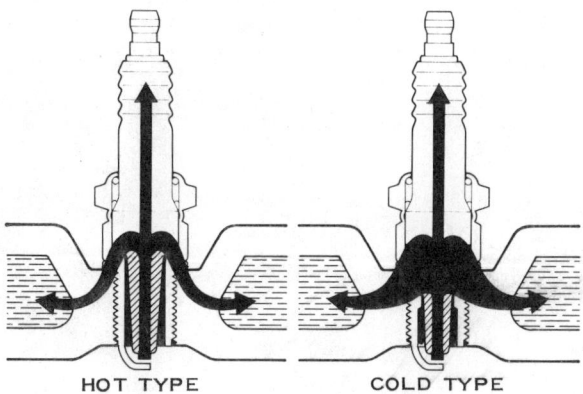

The heat range of a spark plug is determined by the length of the path from the insulator tip to the coolant.

24 GAS GUZZLER'S GUIDE

wire, or the connection inside of the cap may become separated or the boot may be damaged. Remove the spark plugs. **CAUTION: Don't tilt the spark plug socket, or you will crack the spark plug insulator.** Check the spark plugs against the illustrations in the spark plug section of this chapter to determine the operating conditions of the engine. Clean or replace and gap the spark plugs, and then place them on the bench for later installation.

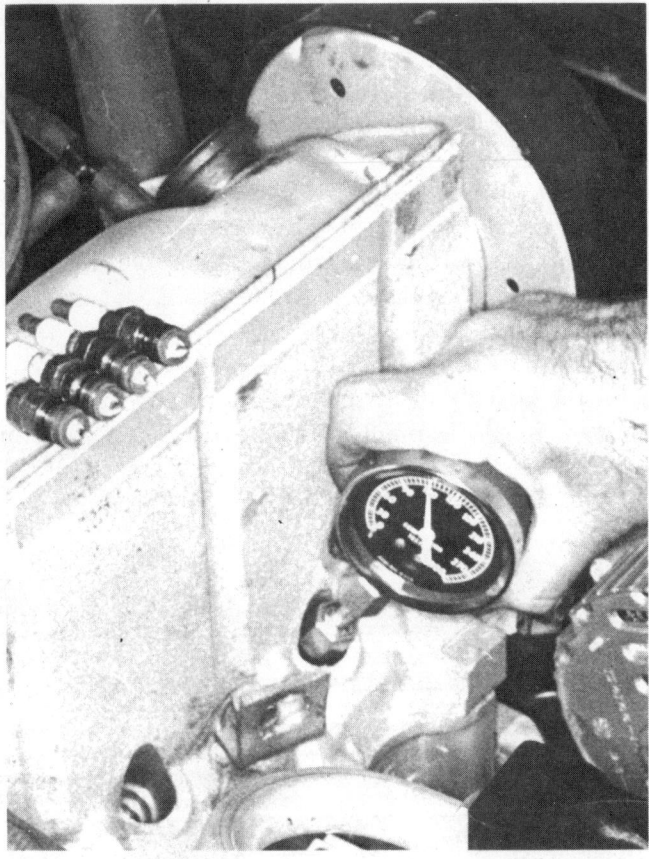

Testing the compression in a Vega engine. Note how the spark plugs are laid out in order of removal so that the firing ends can be read.

The spark plug at the left is normal. The combustion chamber temperatures are high enough to burn off all carbon and oil deposits. An excessively rich air-fuel mixture will deposit soot all over the insulator and shell, right. The combustion temperatures are too low to burn off the deposits. If all of your spark plugs look like this one, you have a defective carburetor.

TESTING THE COMPRESSION

Test the compression by cranking the engine with a gauge inserted in each spark plug hole in turn. The actual reading is not as important as is a variation between cylinders which, if over 20 psi,

An overheated spark plug (left) will show this blistered condition of the insulator. This must have been caused by an excessively lean carburetor air-fuel mixture or an air leak past a defective intake manifold gasket if the condition is localized to two cylinders. A spark plug that is not firing (right) due to the shorting deposits caused by oil and gasoline.

Note the dirty insulator, which should be cleaned with a solvent-soaked rag to prevent flash-over.

indicates ring or valve trouble. Insert a teaspoonful of oil on top of the piston of a cylinder that reads abnormally low, and then crank the engine a few times to distribute the oil. Recheck the compression to see if the oil changed the reading. If the pressure increases, then the compression loss is past the piston rings; if not, the loss is past a burned valve. In either case, you must make the necessary mechanical engine repairs.

SERVICING THE DISTRIBUTOR

To remove the distributor, turn the crankshaft until the rotor points to the front of the engine; mark the position of the rotor on

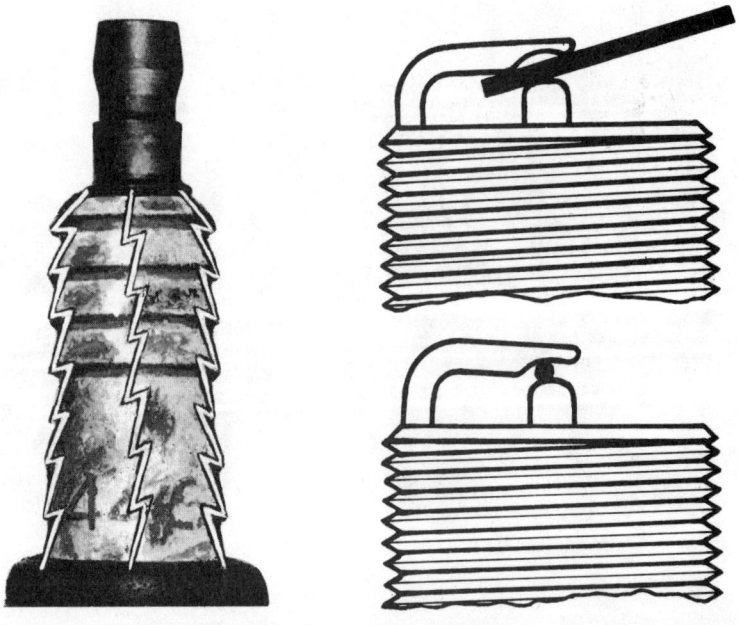

Flash-over occurs (left) when you accelerate hard and the insulator is dirty. The spark jumps over the insulator and the cylinder misfires. A flat feeler gauge will give an erroneous setting (top right) when there is electrode wear. Note how a round gap gauge (bottom right) compensates for this type of wear to give an accurate setting.

Red, brown, yellow, or white colored coatings are by-products of combustion with leaded fuel. This spark plug should be cleaned and gapped to restore it to a good operating condition.

Note the rounded wear on the ground electrode, which is evidence of well over 15,000 miles. Such a spark plug must be replaced to restore good engine performance.

the edge of the distributor housing, and then take out the distributor. **CAUTION: You cannot do a good job of reconditioning or replacing the contact points with the distributor**

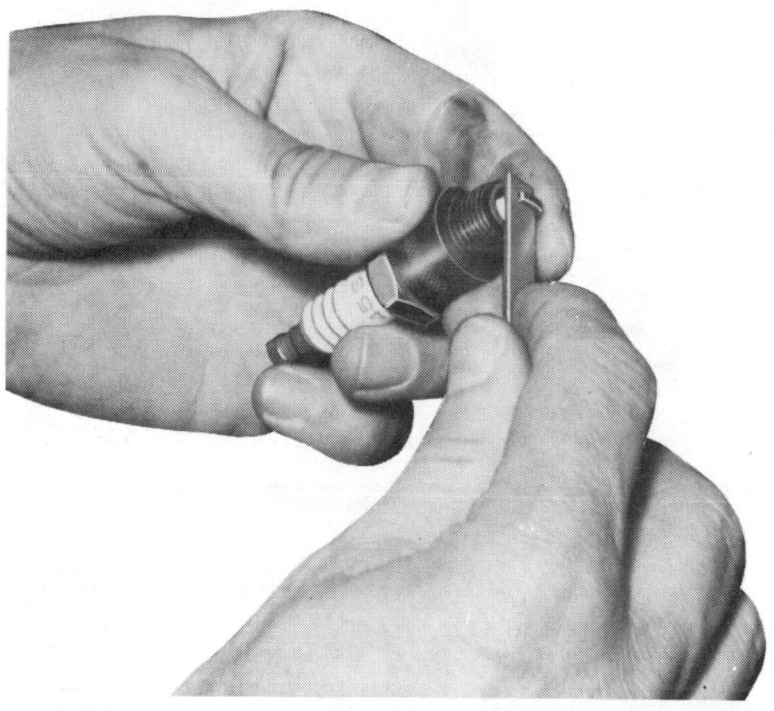

Filing the electrodes square will do more to restore spark plug performance than any other service, if the plug insulator is not shorted out. Be sure to use a file to get the electrodes square. Emery cloth will round them off.

28   GAS GUZZLER'S GUIDE

in the engine. Replace the contact points rather than filing them. It is seldom necessary to replace the condenser. Adjust the point gap to specifications. **CAUTION: Use only a cleaned feeler gauge blade; otherwise, you will coat the contact points with a film of oil, which will cause trouble when it oxidizes.** When making the gap adjustment of a new set of contact points, add 0.003" to the clearance specification to compensate for initial rubbing block wear. *NOTE: It is helpful to keep the contact point retaining screw snug during the adjustment so that the gap does not change when tightening it.*

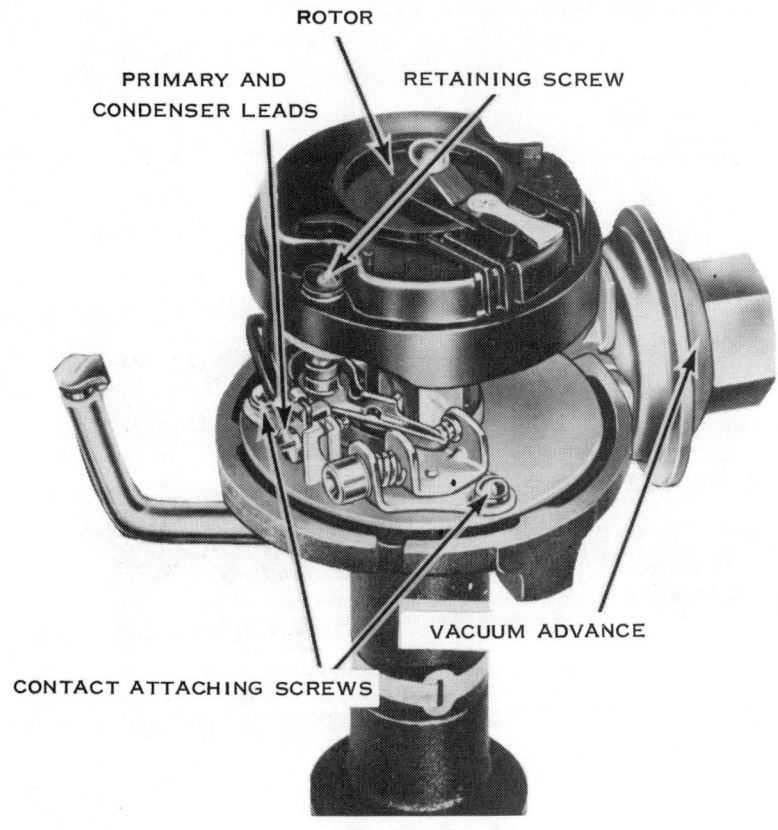

Details of the V-8 engine distributor used on all G.M. engines.

SERVICING THE DISTRIBUTOR 29

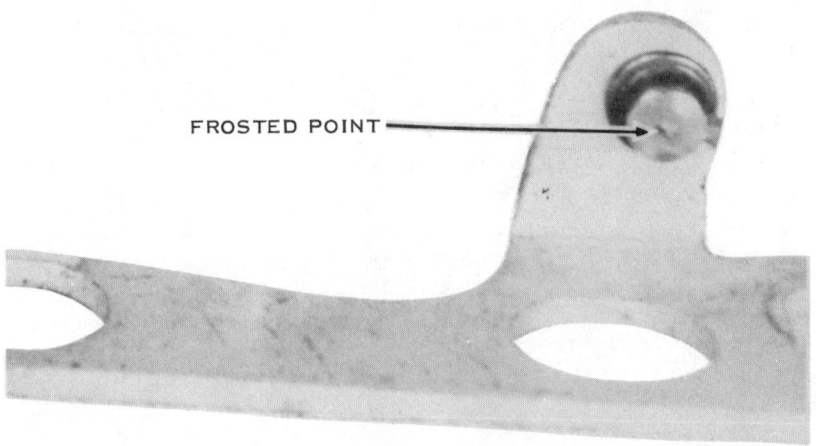

A good contact point has a frosted look. This point set does not need to be replaced.

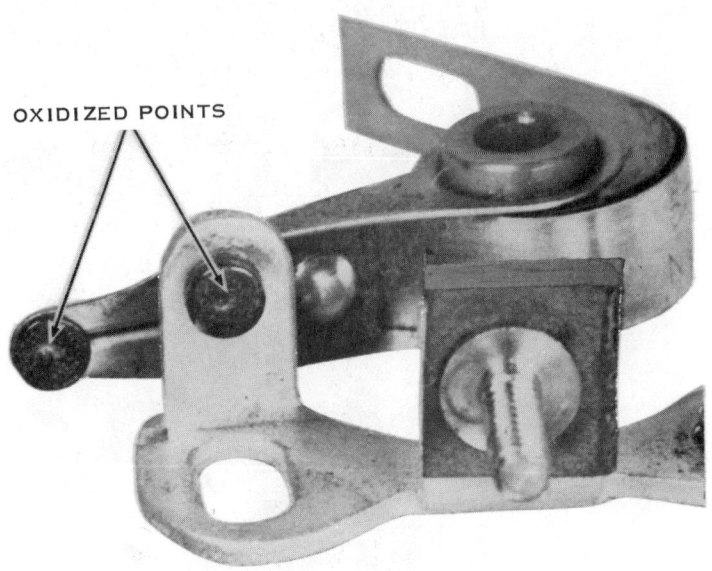

Oxidized contact points which are misfiring badly. In fact they are so bad that they are probably causing starting trouble.

Contact points must be aligned before adjusting the point gap. Always correct misalignment by bending the fixed point support. If you bend the movable point, you will be causing misalignment between the rubbing block and the cam.

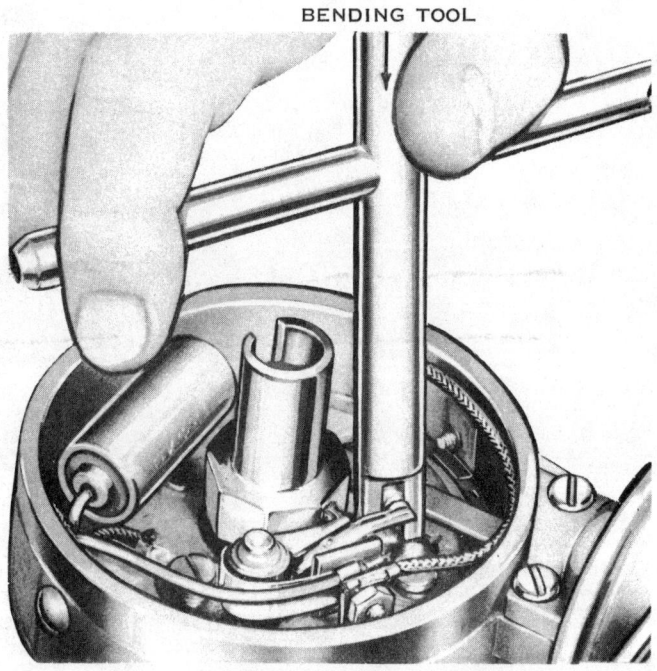

After installing a new set of contact points, make sure that they are aligned properly by bending the fixed contact support. CAUTION: Never bend the breaker arm.

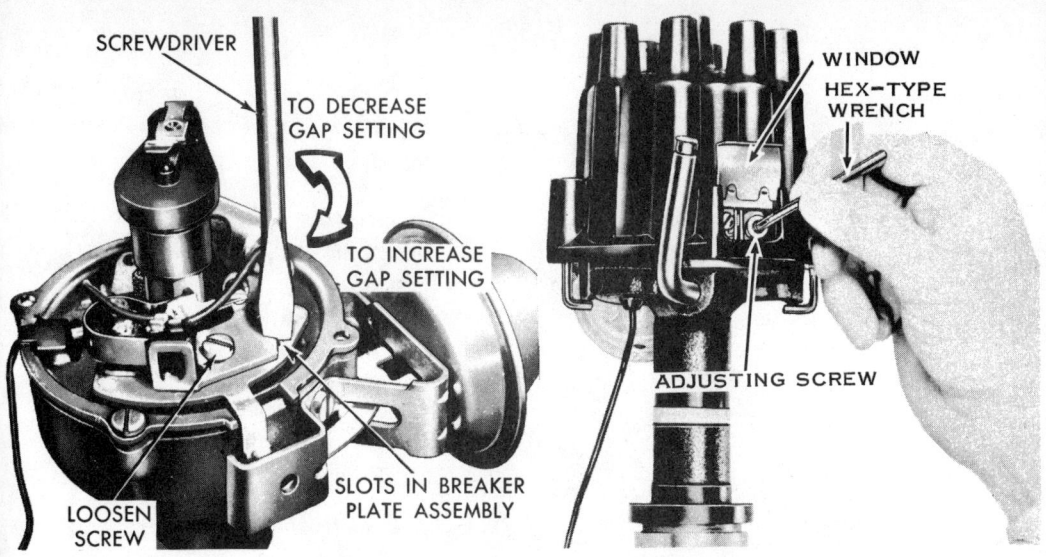

Adjusting the point gap in a G.M. distributor for a six-cylinder engine (left). The late-model G.M. V-8 engine distributors (right) have a window in the distributor cap through which the contact point gap can be adjusted. CAUTION: Always close the window securely after making the adjustment to keep out moisture and dust.

Be careful not to pinch the primary wire under the distributor cap (arrow), or you will have a problem. Take time to be careful.

32 GAS GUZZLER'S GUIDE

Dwell Or Cam Angle

These high-sounding words mean the angle of distributor cam rotation for which the contact points are closed. If you do have a dwell meter, you can set the point dwell rather than measuring the

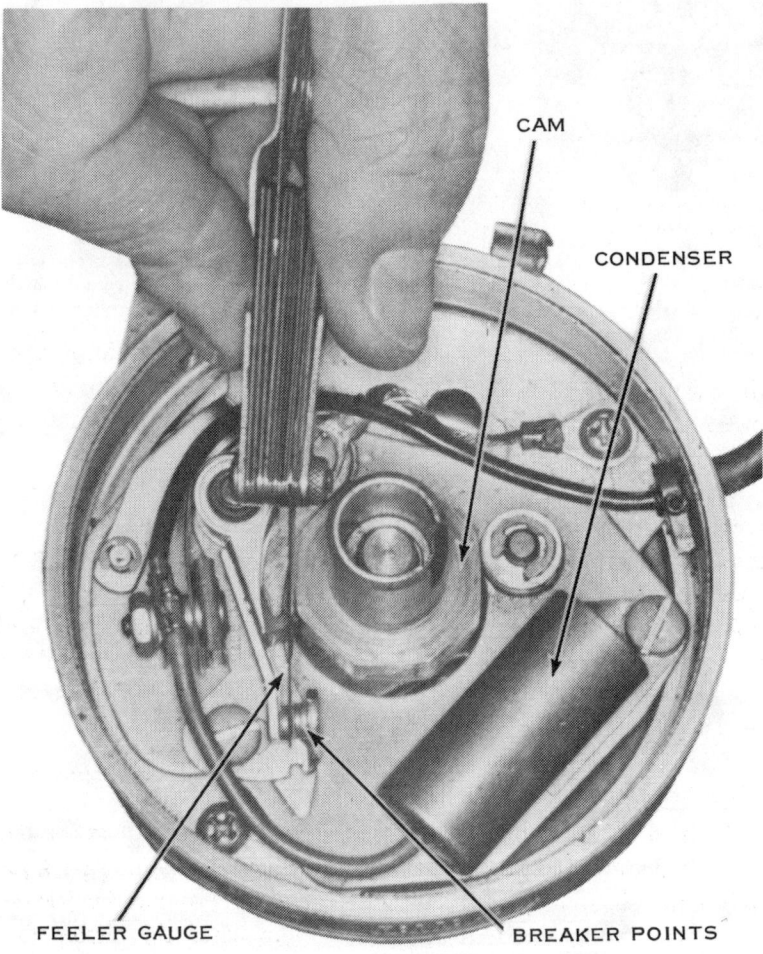

Spacing the contact points on a Chrysler six-cylinder engine. CAUTION: Always use a cleaned feeler gauge to avoid depositing a film of oil on the contact points which will oxidize and cause operating difficulties.

AUTOMATIC ADVANCE MECHANISMS

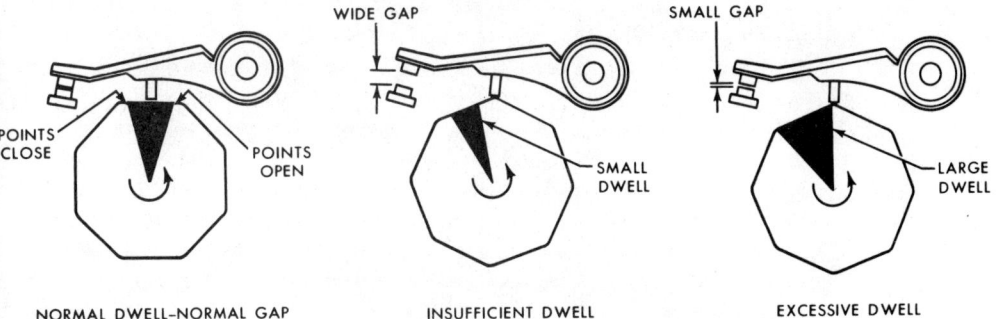

Point dwell is the length of time that the contact points remain together. The right drawings show the effect of wide and narrow gaps on the dwell angle.

spacing with a feeler gauge, as discussed above. If you do, then set the dwell after the distributor is back in place and the engine running. Adjust the dwell to 30° for an eight-cylinder engine, or to 33° for a four- to six-cylinder unit.

Technicians will tell you that you cannot set the dwell accurately with feeler gauges, and that you must have a dwell meter to do this job properly. But all engine manufacturers allow a variation from specifications, usually ±3°, and this means that setting the contact point gap with a feeler gauge can be done just as accurately as with the most expensive tune-up equipment.

Automatic Advance Mechanisms

Centrifugal Advance

Because it takes about 1/350th of a second for complete combustion after the spark plug fires, it is necessary to time the spark just before the piston reaches TDC (top dead center) so that the pressure of the expanding gas will be at maximum when the piston starts down. At higher engine speeds, there is a shorter time interval available for the mixture to ignite, burn, and deliver its power. Consequently, it is necessary to time the spark ealier in the cycle as engine speed increases. This is accomplished by a centrifugal ad-

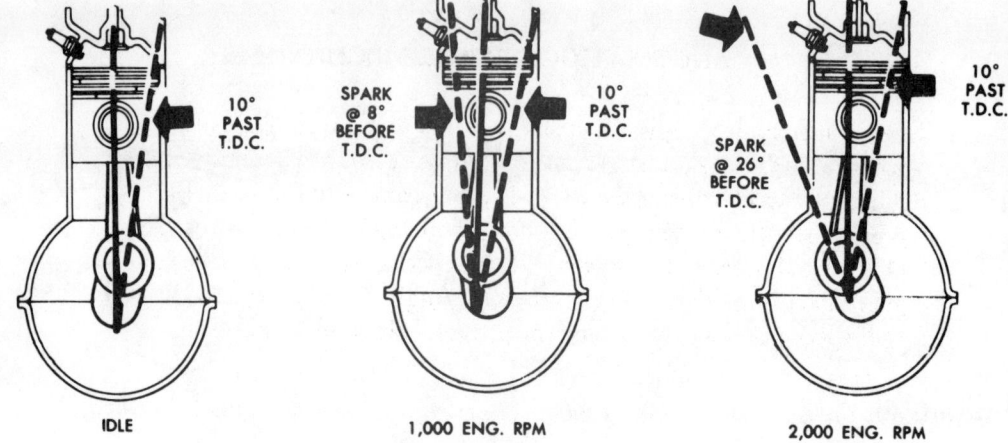

The mechanical spark advance mechanism is designed to advance ignition timing as engine speed increases. At idle speed (left), the spark must occur at TDC for the burning fuel to develop its maximum pressure by 10° ATDC. At 1,000 rpm, it is necessary to advance the ignition timing about 8° for maximum pressure to be developed by 10° ATDC because of the time it takes for combustion to develop. At 2,000 rpm, the spark must occur at approximately 26° BTDC for maximum pressure to develop by 10° ATDC (right).

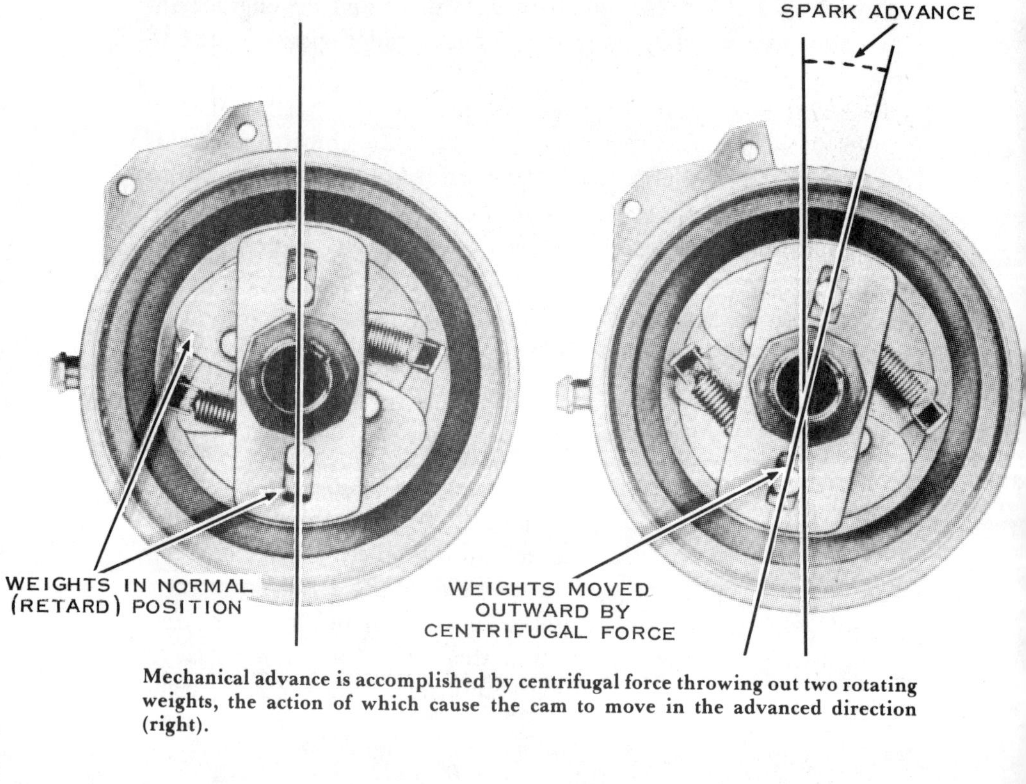

Mechanical advance is accomplished by centrifugal force throwing out two rotating weights, the action of which cause the cam to move in the advanced direction (right).

AUTOMATIC ADVANCE MECHANISMS 35

vance mechanism in the distributor. It contains two control arms which move out against spring tension with increased engine rpm. This motion is transmitted to the breaker cam so that it is advanced with regard to the distributor shaft about 30° crankshaft rotation (15° distributor shaft rotation). If this mechanism sticks or is sluggish in action because of gum, then the engine can be timed right at idle speed, but will not perform properly at higher speeds.

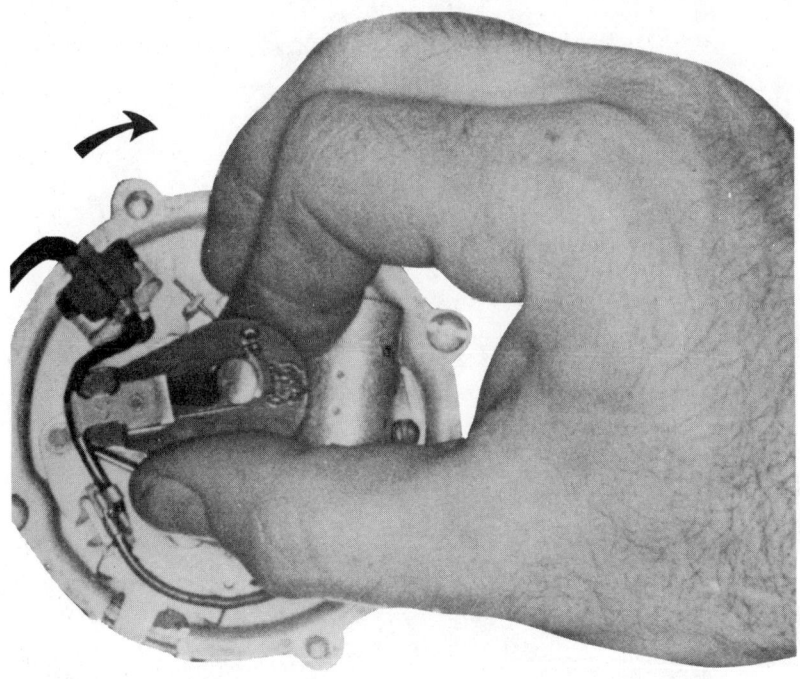

Check the mechanical advance mechanism by turning the rotor. It must feel springy in the normal direction of rotation.

Vacuum Advance

Under light engine loads, the ignition timing can be advanced considerably more than when power is needed and this leads to more efficient engine operation and increased gas mileage. Under heavy loads, however, the ignition timing must be retarded to avoid detonation, which would lead to engine destruction. To provide these timing variations according to load, a spring-loaded diaphragm is connected so as to rotate the entire breaker plate

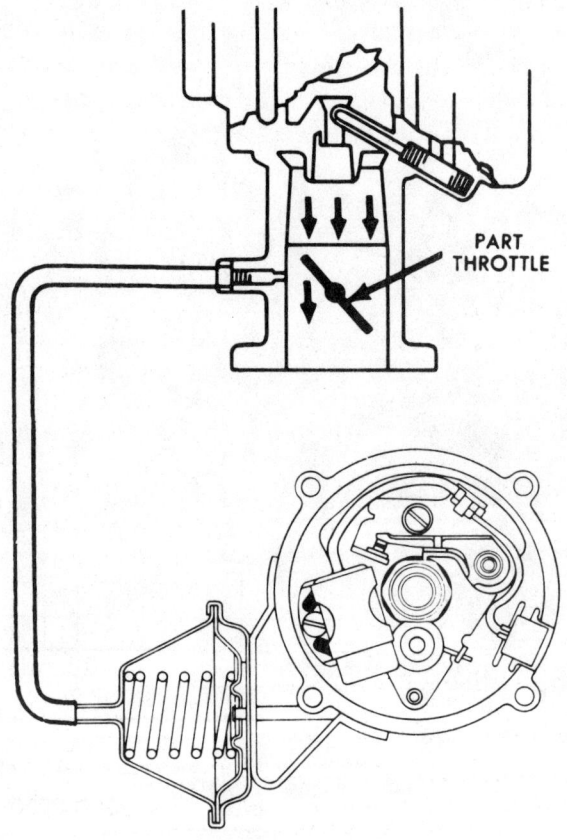

The vacuum-advance mechanism consists of a diaphragm unit which receives its vacuum supply from the manifold side of the throttle valve. Generally, this tap is above the throttle blade so that there is no vacuum to the advance mechanism when the engine is idling. This is especially true on emission-controlled engines.

VACUUM ADVANCE MECHANISM

assembly. The spring-loaded side of the diaphragm is airtight and connected to a vacuum passage in the carburetor, which is on the atmospheric side of the throttle plate when the engine is idling. In this position, there is no spark advance so the engine idles smoothly.

As soon as the throttle moves past the vacuum passage opening (ported spark advance), the vacuum is applied to the airtight chamber so the diaphragm moves against spring tension to rotate the breaker plate to an advanced position.

As the engine is accelerated, intake manifold vacuum drops sharply, and the vacuum-advance unit retards the ignition timing to minimize detonation, which would otherwise occur with advanced ignition timing and high combustion chamber pressures.

The total timing advance of an engine, therefore, is regulated by the positions of both the vacuum advance and the centrifugal advance mechanisms.

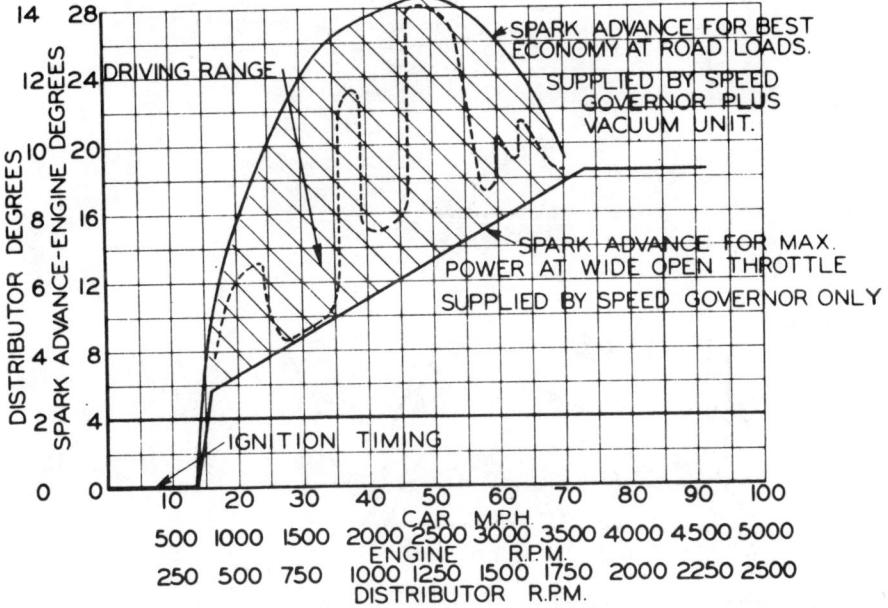

In operation, the ignition timing can be at any point in the shaded section of this graph due to the action of the centrifugal advance mechanism, modified by the action of the vacuum-advance unit.

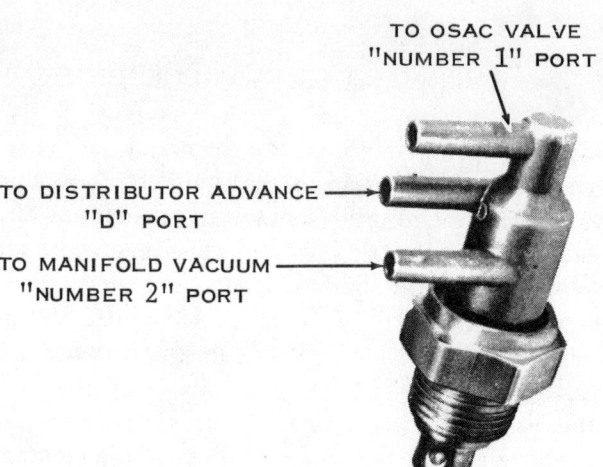

TO OSAC VALVE "NUMBER 1" PORT

TO DISTRIBUTOR ADVANCE "D" PORT

TO MANIFOLD VACUUM "NUMBER 2" PORT

The Thermal Vacuum-Switching (TVS) valve is designed to advance the ignition timing of an idling engine that is overheating.

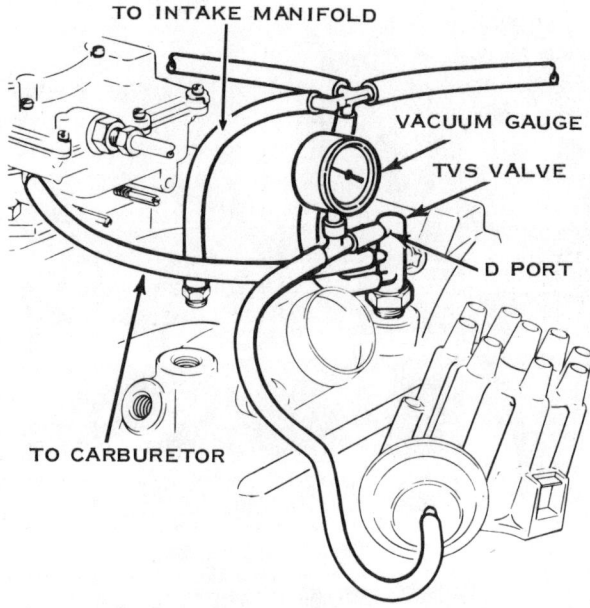

TO INTAKE MANIFOLD

VACUUM GAUGE

TVS VALVE

D PORT

TO CARBURETOR

Hose and vacuum gauge hookup for testing a TVS valve. Restrict the incoming air to the radiator with a piece of cardboard and let the engine idle. The vacuum gauge should show an 18″Hg reading when the coolant temperature reaches 230°F., which indicates that the vacuum-switching valve has changed the vacuum source from a ported carburetor one to a direct intake manifold tap.

Thermal Vacuum-Switching Valve

A thermal-sensitive valve is used on most of today's emission-controlled engines to advance ignition timing on an idling engine that is beginning to overheat. The valve switches the vacuum source to the distributor vacuum-advance unit from a ported one to a direct intake manifold tap, and this can advance the ignition timing of the idling engine about 15°, which will increase engine idling speed and coolant flow.

While this valve is used basically for engine cooling purposes, it is mentioned to show the reader the devices added to modern engines.

A leaking or torn vacuum-advance actuator will cause a loss of gas mileage during light-throttle operation.

40   GAS GUZZLER'S GUIDE

If the vacuum hoses are hooked up improperly, engine performance will be affected. In fact, any defect in the vacuum system will even affect the shifting of the automatic transmission.

Servicing The Automatic Advance Mechanisms

Before replacing the distributor, make sure that the advance mechanisms are functioning properly. *To test the centrifugal advance mechanism,* twist the rotor and it must feel springy when rotated in one direction and solid in the other. *NOTE: If you have the distributor out of the engine, you will have to hold the bottom end of the shaft in a vise to make this test.* If the rotor is not springy, the advance mechanism is stuck; it can be gummed or rusted. In either case, you must remove the breaker plate to lubricate and free up the mechanism. If the rotor is springy in one direction (function-

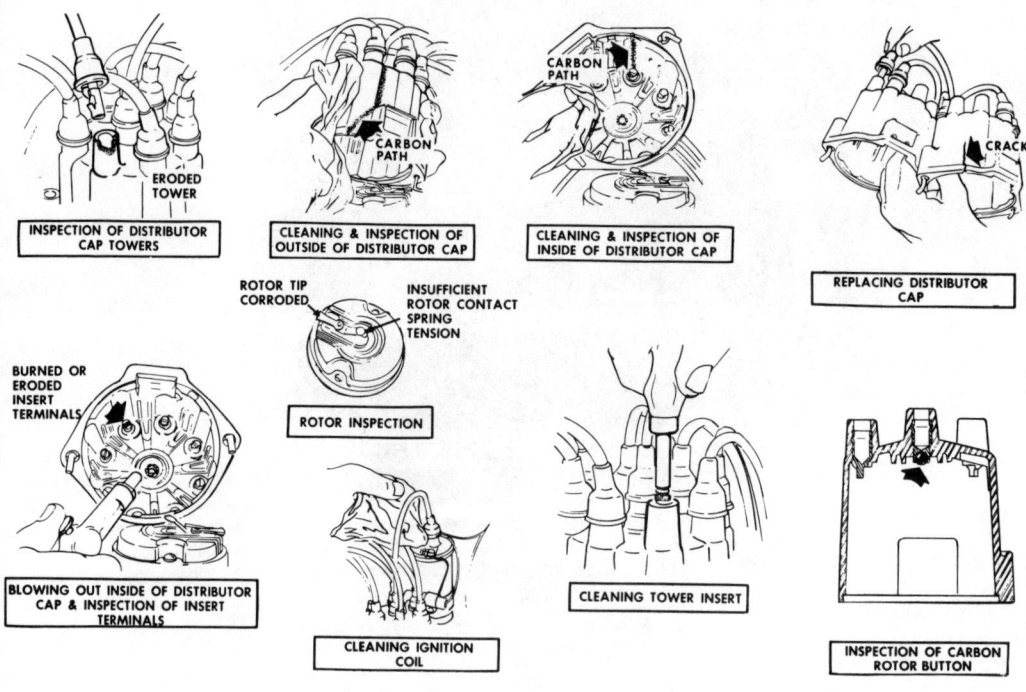

Cleaning and inspecting the distributor cap.

REPLACING THE DISTRIBUTOR

ing properly), place one drop of engine oil on the wick in the center of the shaft.

To check the vacuum advance mechanism, apply oral vacuum to the fitting at the end of the unit, and you should see the breaker plate rotate. Now, hold the tip of your tongue over the fitting to retain the vacuum, and the plate must not move back; otherwise, the diaphragm is porous and the unit must be replaced.

Replacing The Distributor

After making the gap adjustment, apply a light layer of heavy grease to coat the distributor cam. Turn the distributor shaft in the

Check the inside of the distributor cap for a crack between terminals, which will cause engine misfire and loss of mileage.

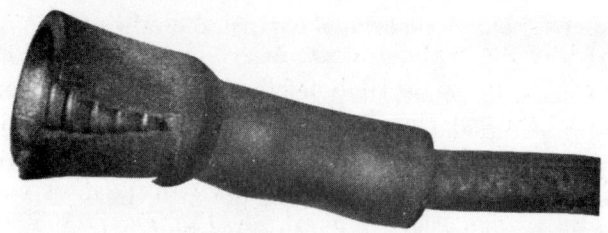

Check the condition of the high-tension wires to make sure that they are not old and brittle, which will cause engine misfiring and hard starting.

Dirt and grime on the surface of the ignition coil can cause high-tension leakage, which will result in misfiring and hard starting, especially in wet weather.

SETTING THE IGNITION TIMING 43

normal direction of rotation so that the lubricant is wiped off against the back of the rubbing block, where it remains as a reservoir to supply lubricant as the rubbing block wears. Wipe the excess lubricant from the cam, leaving only the grease stored behind the rubbing block. Replace the distributor in the engine, with the rotor pointing toward the front and aligned with the mark you made on the housing edge before removal. *NOTE: The ignition timing will be adjusted after the engine is running.*

Replace the cleaned and gapped spark plugs. Tighten them to specifications. Start the engine and allow it to warm to operating temperature. If the tappets are mechanical, adjust the valve lash to specifications at this time.

SETTING THE IGNITION TIMING

Adjust the engine idle speed to the rpm specified on the tuning decal. **CAUTION: The contact point gap (or dwell) must have been set first, because it affects the timing.** Connect a timing light to an adapter for No. 1 spark plug. **CAUTION: Don't puncture the high-tension wire, or you will damage the core.** Disconnect the vacuum line(s) to the distributor and plug the source(s). Point the timing light toward the timing indicator. The specified timing mark should line up with the pointer. If it doesn't,

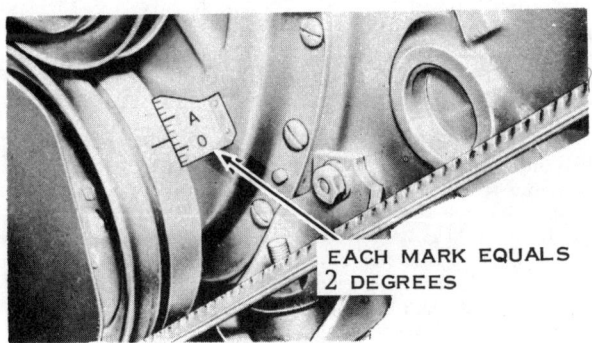

Timing indicator plate, with each mark equaling 2°. "O" is TDC and "A" means advanced (Before Top Dead Center).

Typical high-tension wiring. Note the clips used to keep the wires separated so that cross-firing does not occur.

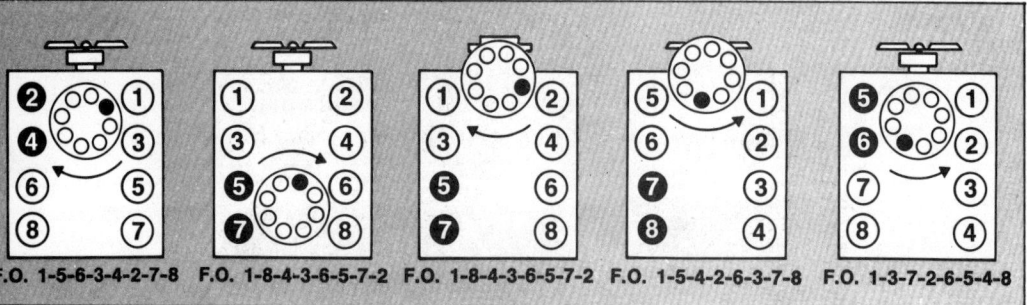

F.O. 1-5-6-3-4-2-7-8   F.O. 1-8-4-3-6-5-7-2   F.O. 1-8-4-3-6-5-7-2   F.O. 1-5-4-2-6-3-7-8   F.O. 1-3-7-2-6-5-4-8

Ignition cross-fire is a condition in which a spark plug fires out of time through a process called induction. This can cause engine roughness, backfiring, detonation, and serious engine damage. An electrical field develops around a spark plug wire when high voltage travels through it on the way to the spark plug. There is sufficient energy in this field to induce voltage in an adjacent wire if: (1) the adjacent wire is close enough and runs parallel to it, and (2) the wire runs to an adjacent cylinder which is next in the firing order. The result is that both fire, but one is advanced. To avoid this, spark plug wires must be installed in their original order. Above all, never tape the wires together in a neat bundle as this will aggravate this condition. These diagrams illustrate the firing orders (F.O.) of most V-8 engines. The cylinders in black are the critical ones and must be kept well separated to prevent crossfire.

loosen the distributor hold-down bolt and rotate the distributor until the mark lines up with the pointer. Tighten the hold-down bolt and recheck the timing. Connect the distributor vacuum line(s). Now, accelerate the engine to see if the centrifugal advance mechanism is operating. The position of the mark should advance on the pulley if the unit is in good condition. Reconnect the vacuum line(s).

POWER-TIMING THE ENGINE

Purists, especially those with elaborate tune-up equipment, will shake their heads in wonder at this one, but it is quite possible to adjust the ignition timing for optimum results with a vacuum gauge and a road test. They will point to the tuning decal and say that the specifications are, for example, exactly 5° BTDC (Before Top Dead

This illustration shows the high-tension wiring on a late-model Ford engine, with a firing order of 1-5-4-2-6-3-7-8. Cylinders 7 and 8 fire one after the other. Note how Ford has arranged the wires so that they're well separated to prevent crossfire.

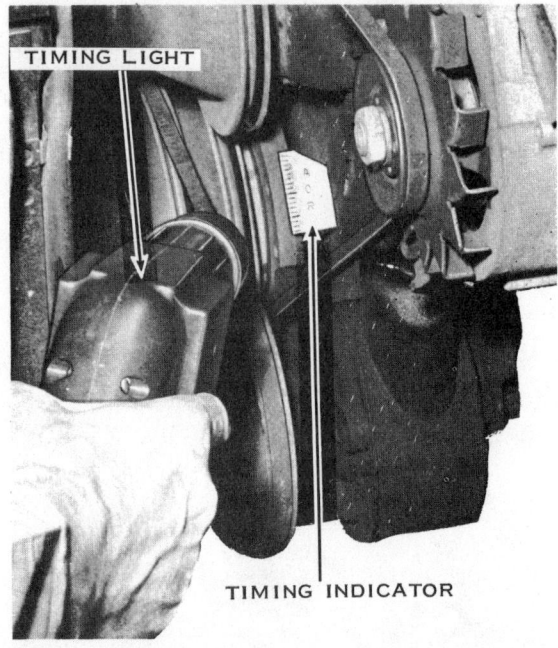

Use a timing lamp to set the ignition timing to specifications, as discussed in the text.

Preignition is premature ignition of the fuel charge and this can be caused by advanced ignition timing or any hot spot within the combustion chamber capable of initiating ignition of the remaining fuel charge.

POWER-TIMING THE ENGINE 47

Center), but the manufacturer always gives a leeway of plus or minus 2 or 2-1/2° from that **"exact"** figure, which means that the engine in question can be timed anywhere from 7-1/2° BTDC to 2-1/2°BTDC and still be within specifications, this applies to all modern emission-controlled engines, as well.

To prove that the above is true, the following is a quote from Chrysler Technical Service Bulletin No. 08-09-74D, which is entitled, *A method of reducing hydrocarbon emissions*. "The basic ignition timing of California passenger car models with 360-4 Bbl. standard-performance and 360-4 Bbl. high-performance engines has been changed from 5° BTDC to 2-1/2° BTDC." By advancing the ignition timing from that specified on the tuning decal,

Preignition can cause the top of the piston to melt.

hydrocarbon emissions were reduced and, **most important to our readers**, the performance and gas mileage of this engine was increased significantly.

To power-time an engine, connect a vacuum gauge to any tap on the intake manifold, start the engine, and then open the throttle enough to engage the fast-idle cam to achieve a fast-idle speed of about 2,000 rpm. *NOTE: The exact speed is not important.* Now, turn the distributor body slowly to obtain the highest vacuum reading on the gauge, and then back it off slowly until the needle dips slightly from the maximum reading. The exact amount of backing off needed must be determined by the engine itself. If there is a lot of carbon in the combustion chambers or you are using a low-test gasoline, you will have to retard the timing more than for a clean engine running on leaded fuel. *NOTE: The more advanced you can run the ignition timing, the better gas mileage you will obtain.*

Now, road-test the car to determine if the engine pings. If it does, retard the timing a small amount and repeat the road test. Do this until you lose all traces of ping. Now your engine is timed to the most efficient running position for its condition.

CAUTION: Don't run the engine with the timing too far advanced or the preignition will destroy the spark plugs, valves, and pistons.

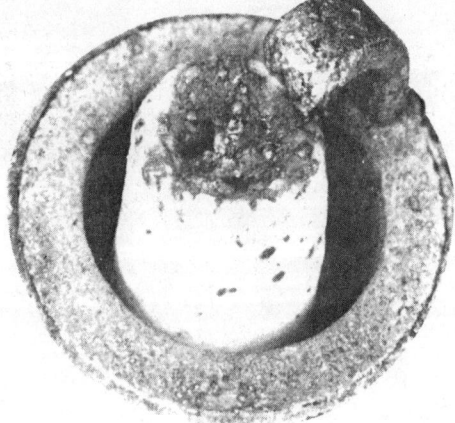

A spark plug that is exposed to preignition will have a blistered porcelain.

SERVICING THE FUEL SYSTEM   49

FUEL SYSTEM SERVICE

Carburetor repairs and adjustments must be made only after having serviced the distributor and adjusted the ignition timing, because these adjustments affect carburetion to a large extent. Before attempting to adjust the carburetor, it is good practice to clean or replace the air cleaner element and fuel filter. A partially clogged air filter element lowers your gas mileage to a considerable degree, depending on the amount of accumulated dirt. The fuel filter should be changed every 10,000 miles to keep small particles from passing into the fuel bowl, which can sometimes hold the car-

Water damage inside of the fuel pump. The same condition occurs inside of the carburetor.

50 GAS GUZZLER'S GUIDE

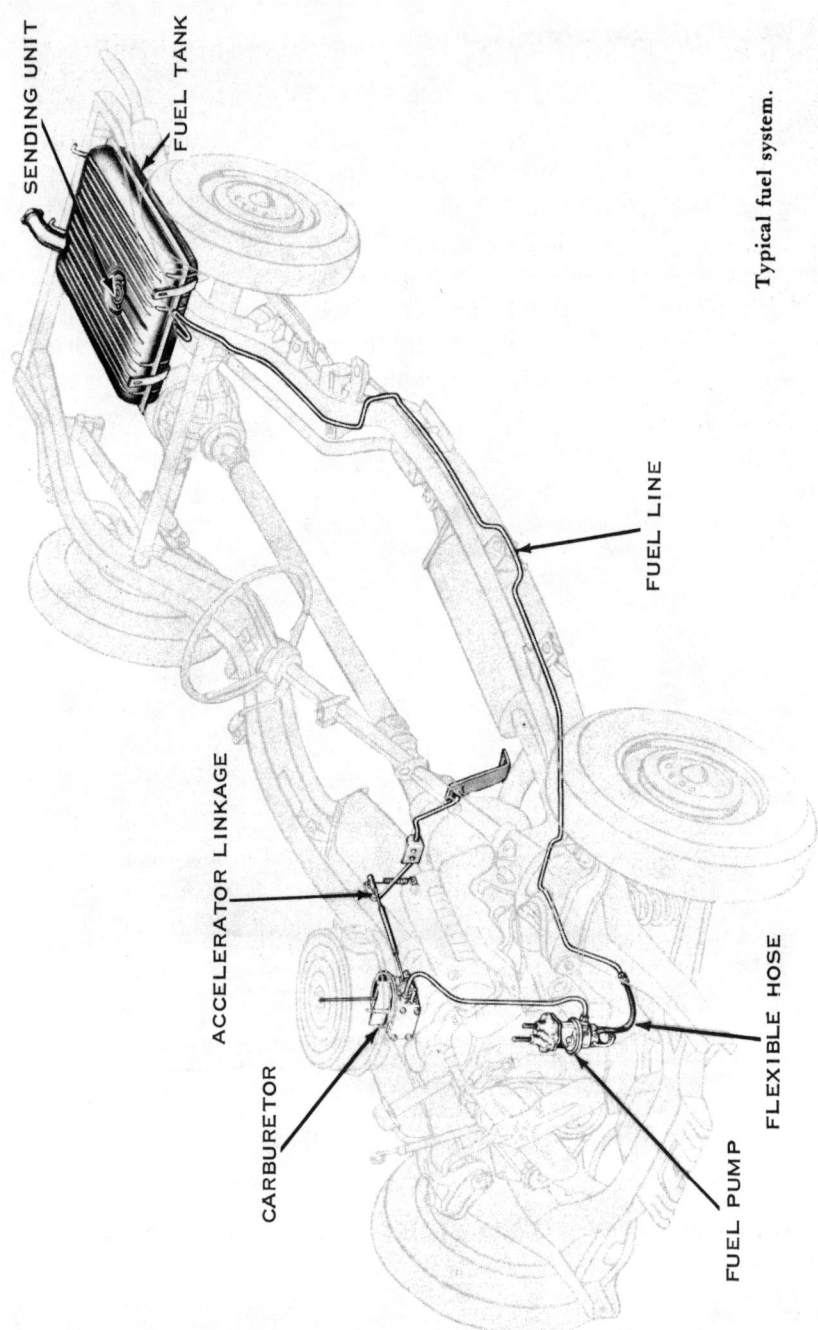

Typical fuel system.

SERVICING THE FUEL SYSTEM 51

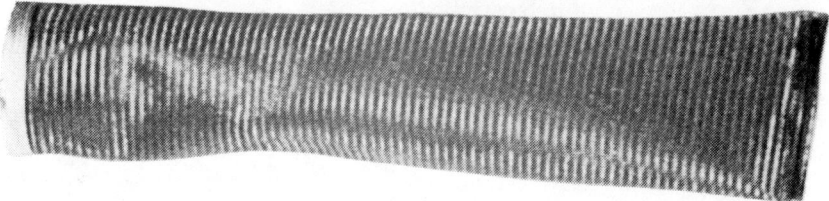

If you do have water in the gas tank, then you will find this type of strainer at the end of the intake pipe, inside of the tank, plugged enough to cause operating difficulties. The remedy is to remove the tank, drain and flush it, and then replace the strainer.

buretor needle valve off its seat and cause flooding; this will reduce gas mileage. These small service jobs can be performed by anyone, and they sometimes significantly reduce your operating costs, usually way above the purchase price of the filters. And, too, they

The metal cover of the fuel pump shows this kind of damage from water in the gasoline.

52   GAS GUZZLER'S GUIDE

give you more trouble-free mileage.

If your carburetor needs to be serviced, then this work must be done before making the final adjustments. Carburetor repairs are called for when your mileage and engine performance drop off suddenly and you can tell from the spark plugs that the mixture is too rich or lean.

AIR FILTERS

For every gallon of gasoline passing through the carburetor, it

Defective fuel pump diaphragm shows how the constant flexing causes the diaphragm to crack, and this can cause an internal fuel leak into the crankcase, which will lower your gas mileage considerably. The remedy is to replace the fuel pump.

SERVICING THE AIR FILTER 53

requires as much as 9,000 gallons of air for a combustible mixture. All the air must pass through the filter element in the air cleaner and in 10,000 miles of operation, the filter will start to plug up with dirt; in 24,000 miles it can be costing about $46.00 a year in lost gas mileage. In addition, the rich fuel mixture will be destroying the cylinder wall lubrication to cause excessive engine wear, and it will also be raising your HC and CO emissions above legal limits.

The answer is to clean or replace the filter element periodically. This is one of those simple service operations that anyone can perform, and you're home cost-free in addition.

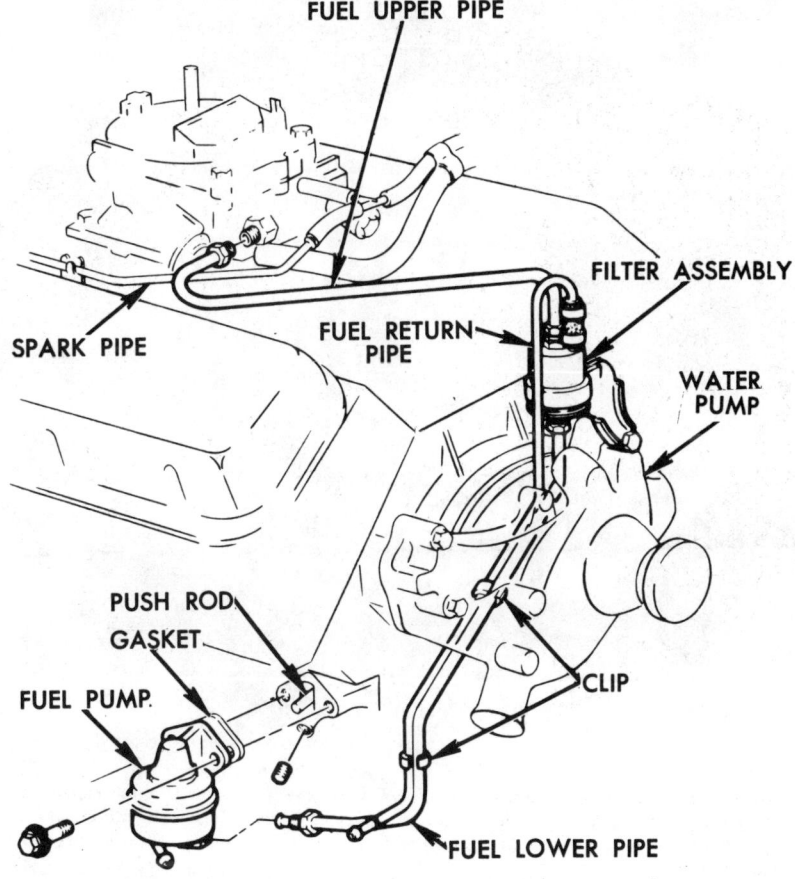

An in-line fuel filter should be changed every 12,000 miles.

A clogged air filter element will enrich the air-fuel mixture and lower your gas mileage proportionally.

The fuel filter element in the carburetor bowl should be changed every 12,000 miles.

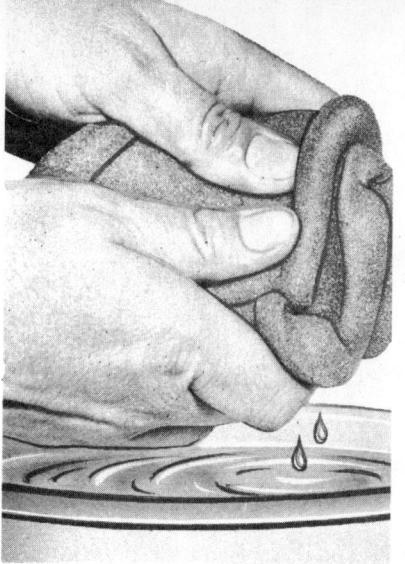

Cleaning a polyurethane filter element. Squeeze it dry after soaking it in solvent. CAUTION: Don't wring it out, or you will tear it.

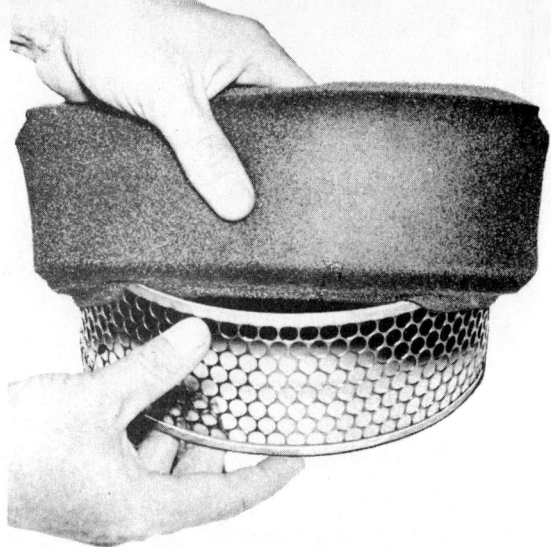

Some filter elements are made of polyurethane. They have to be supported on a mesh element. Be careful to seat the filter all around the support after it is cleaned.

Compare this new filter with the previous one.

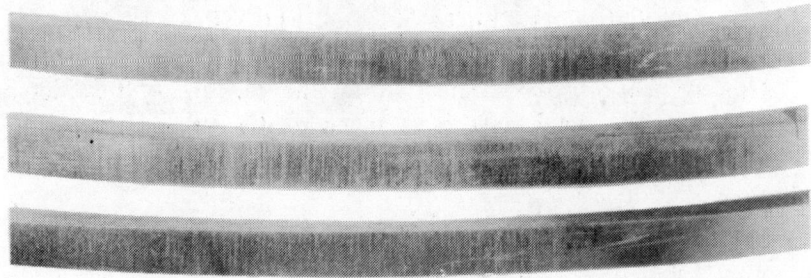

Surface scratches on a set of piston rings show how the dust in unfiltered air damages the moving parts of an engine.

A dry filter element can be cleaned by directing a low-pressure air flow through the element in a reverse direction. Hold the filter up to the light to be sure that there are no holes to admit unfiltered air.

THERMOSTATICALLY-CONTROLLED AIR CLEANER

This system provides heated air to the carburetor induction system. A sheet metal stove is attached to the exhaust manifold

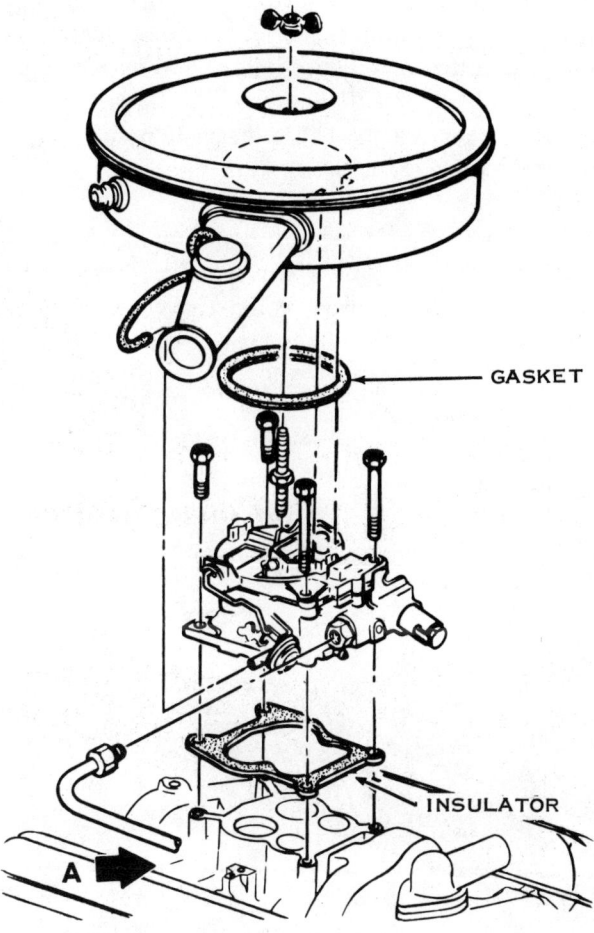

When replacing the carburetor, always tighten the hold-down bolts evenly to compress the insulator properly for avoiding an air leak. CAUTION: Uneven tightening can crack off one of the carburetor ears. CAUTION: Always replace the gasket between the air cleaner and the carburetor to make sure that no unfiltered air enters the engine to cause damage to the moving parts.

where underhood air is heated as it passes over the hot exhaust manifold. The heated air is conducted from the stove to the air cleaner through a flexible duct.

The air cleaner is designed to control the inducted air temperature at approximately 100°F. through about 70 mph road-load conditions. At speeds above 70 mph, decreasing intake manifold vacuum and an increasing differential pressure across the temperature-controlled door cause the door to open gradually until an intake manifold vacuum of about 5.5″ Hg is reached; at this time the heat-control door will be in the heat-off position, and the inducted air temperature will be the same as underhood air temperatures.

The use of a heated-air system does not materially affect the inducted air temperature during warm weather, but it does quickly raise the intake air temperature in cold weather. A decreased spread in the temperature range permits the use of leaner air-fuel mixtures with satisfactory driveability.

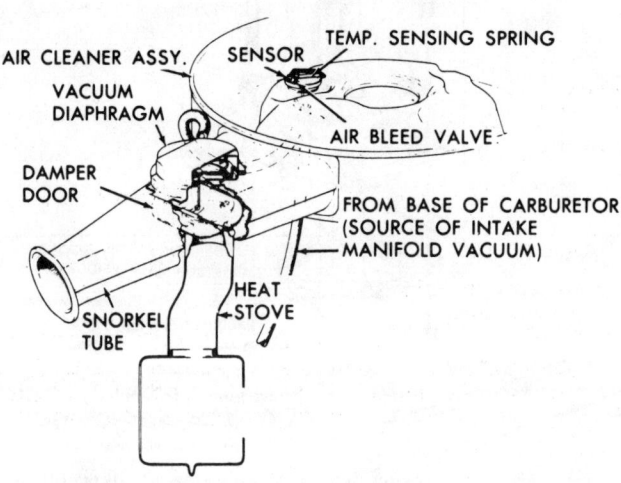

Modern emission-controlled engines have thermostatiscally-controlled air cleaners, as discussed in the text.

GAS MILEAGE IMPROVERS 59

SERVICE PROCEDURES

Check to see that the cold-air door is open before the engine is started, and that it closes immediately after starting as vacuum is built up. Then the door should open again gradually as the engine warms to operating temperature. A thermometer can be used to check the temperature of the thermostat, and this should begin to control the vacuum motor for starting to close the heated-air door at about 105°F.

GAS MILEAGE IMPROVERS

There are a great number of "add-on" devices advertised in newspapers and auto enthusiasts' magazines that promise large increases in gas mileage, and sometimes they also promise increased power as well as easier starting. While the author has not had the opportunity (pleasure) of testing each of the advertised devices, and one of them might just fulfill the advertised claims, they are, as a

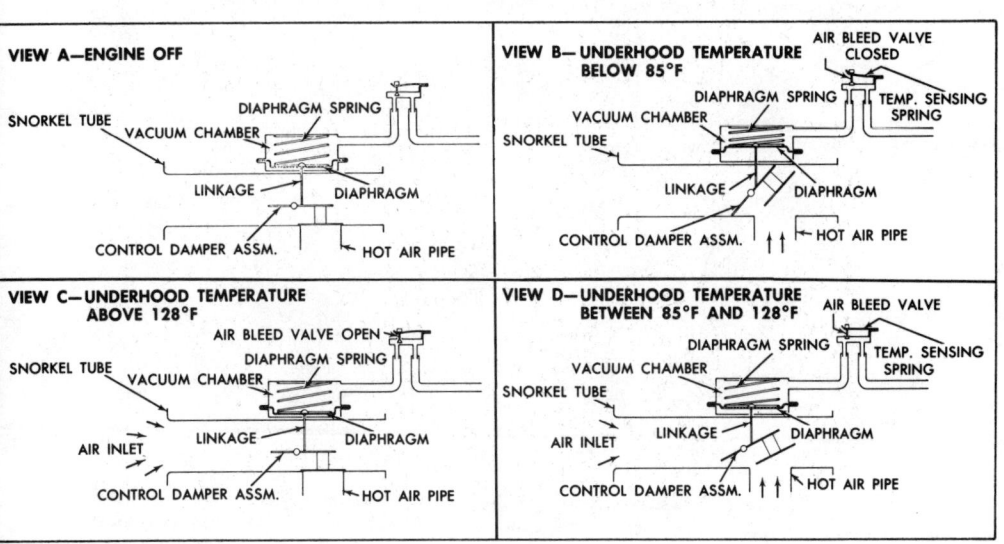

Operation of the thermostatically-controlled air cleaner.

60 GAS GUZZLER'S GUIDE

 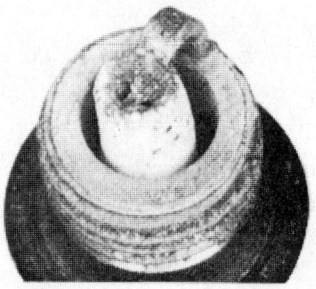

Preignition damage to a piston (left) and a spark plug.

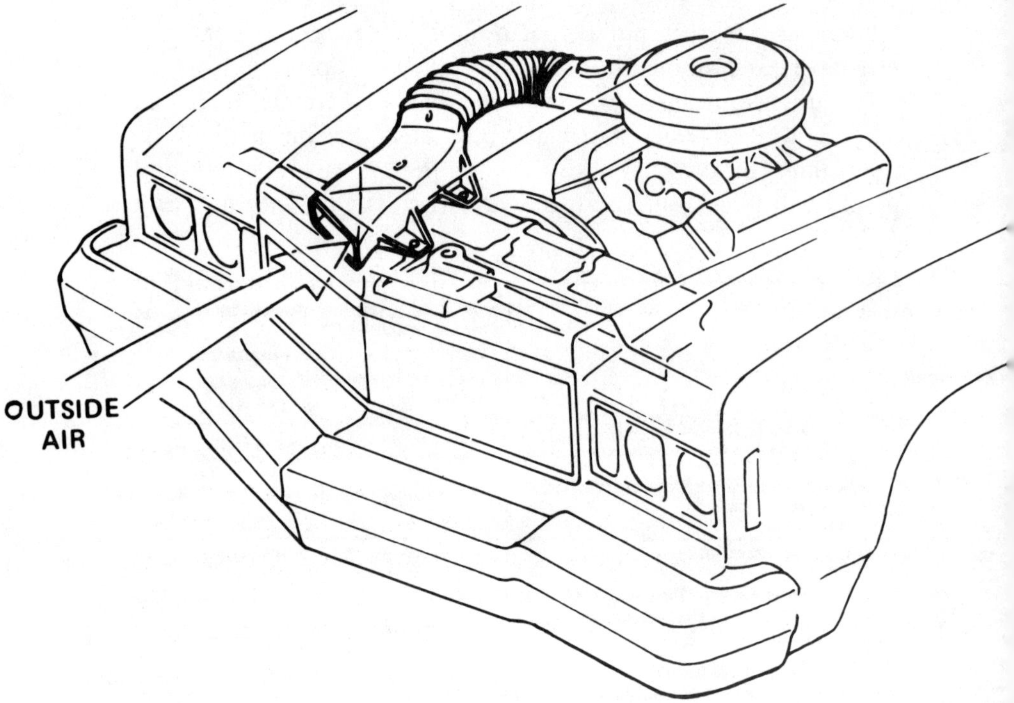

OUTSIDE AIR

This drawing shows how the 1975 Chevrolet engine takes outside air into the air cleaner to improve engine performance. The cooler, outside air is denser than the warmed underhood air, consequently the cylinders are charged with more molecules of usable air-fuel mixture. The result is improved performance throughout the engine speed range. You can manufacture such a kit and increase your gas mileage at least 10%.

group, completely worthless for increasing your gas mileage, UNLESS your carburetor air-fuel mixture is extremely rich (too much fuel being mixed with the air).

Most of these accessories are air bleeds into the intake manifold, and some of them have special so-called ultra-sonic vibrator units for the advertised purpose of breaking up the molecules of fuel so that a better mixture results—so the claims say. However, none of these devices that the author has tested increased the gas mileage one iota in an engine where the carburetor was delivering the correct air-fuel ratio.

In modern emission-controlled engines, where the air-fuel mixture is extremely lean to start with in order to meet emission standards, the addition of an extra air bleed can result in an excessively lean mixture, detonation, and burned valves. In extreme cases, it can result in valve head breakage, which can destroy the engine.

If such a device does increase the gas mileage, then the carburetor air-fuel mixture is too rich and the carburetor should be overhauled. It must be worn to the point where the "timing" of jet opening must be incorrect because of worn linkage, and the float valve and seat must be worn to the point of raising the float level above specifications, leaking enough to allow excess fuel to enter the carburetor float bowl. Both of these conditions will cause the fuel level within the float bowl of the carburetor to rise above specifications, and the enriched mixture will be evident at all speeds and operating ranges, destroying the performance-effectiveness of the engine.

There is a way, however, to increase gas mileage and performance in your engine without endangering it. Most engines draw heated underhood air into the air cleaner, and this air has less molecules per square inch (it is thinner) than does cooler, outside air. Many modern high-performance engines are ducting outside air to the air cleaner, and the resulting denser air (more molecules) increase the volumetric efficiency (big word, meaning the ability of the engine to breathe better), and this means better gas mileage and performance. If Detroit engineers are doing it, you can rest assured that you can do it safely, too. A piece of flexible tubing, fittings, and some old-fashioned American ingenuity should do the trick.

CARBURETORS

The carburetor furnishes a correctly proportioned air-fuel mixture to be burned in the combustion chambers. Dirt and gum restrict the flow of fuel, causing a lean operating condition; hesitation on acceleration results. Gum and carbon form in the automatic choke mechanism, resulting in the choke being applied for a longer than normal time, lowering the gasoline mileage proportionally. Wear occurs in the linkage, changing the timing of the mixture, which results in poor operation and lowered fuel mileage. The remedy, of course, is to clean and overhaul the carburetor and to make the adjustments which will restore it to its former operating efficiency.

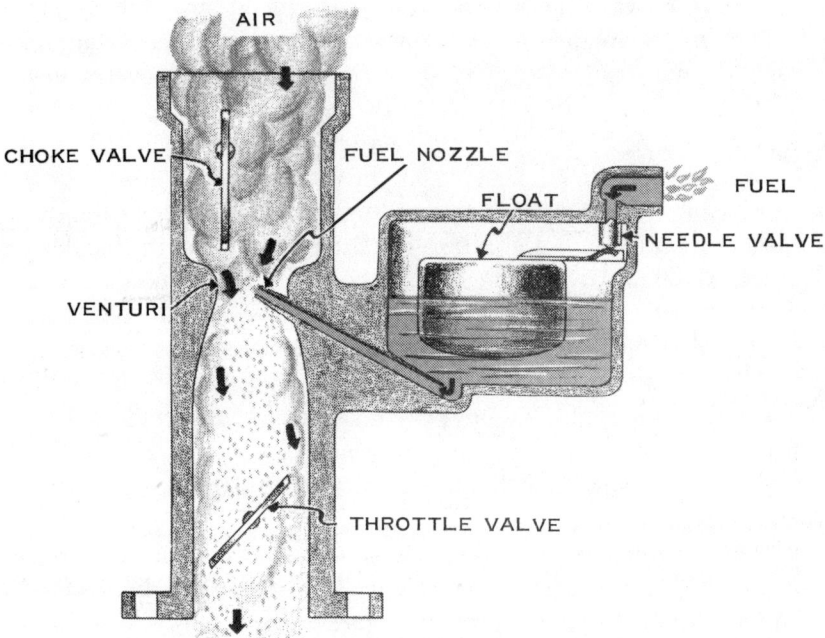

Sectioned view through a carburetor to show the passage of air and fuel. The throttle valve is partially opened in this drawing so that the engine is running at a normal road speed.

CARBURETORS 63

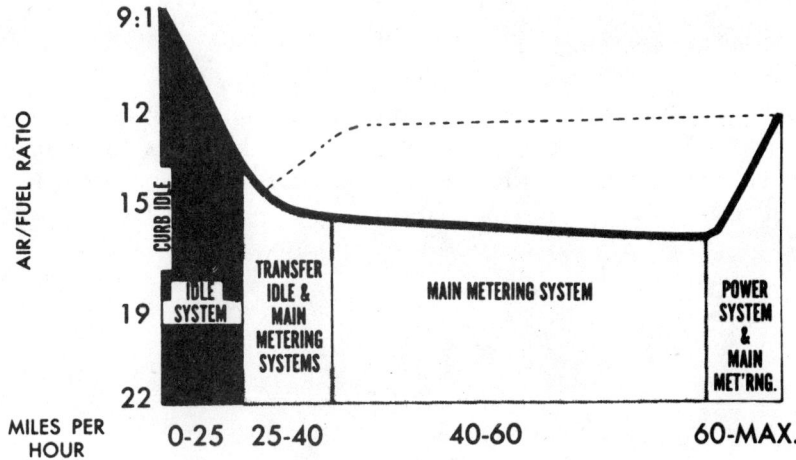

Carburetor flow curve showing the various systems in operation.

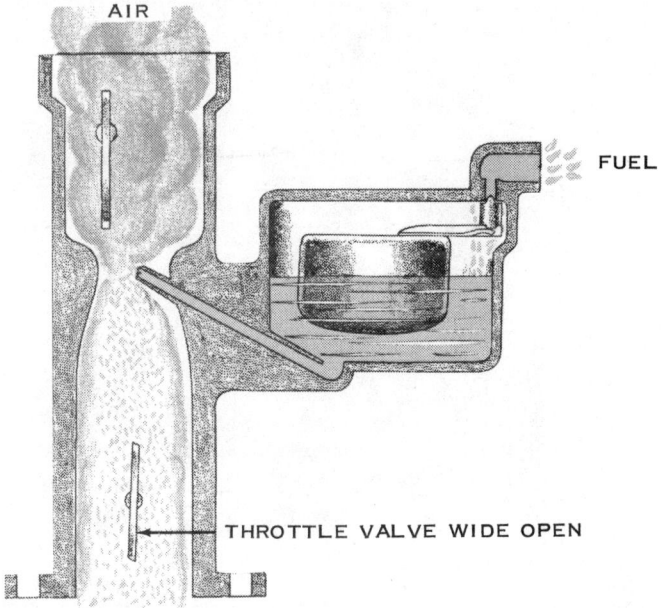

In this diagram, the throttle valve is wide open, and the engine is running at top speed.

SERVICE PROCEDURES

Because of the great number of carburetor types, it would be impossible in this kind of book to provide overhaul instructions for all of them. Instead this section will include a number of hints for replacing worn carburetor parts and for making adjustments that will restore your carburetor's efficiency.

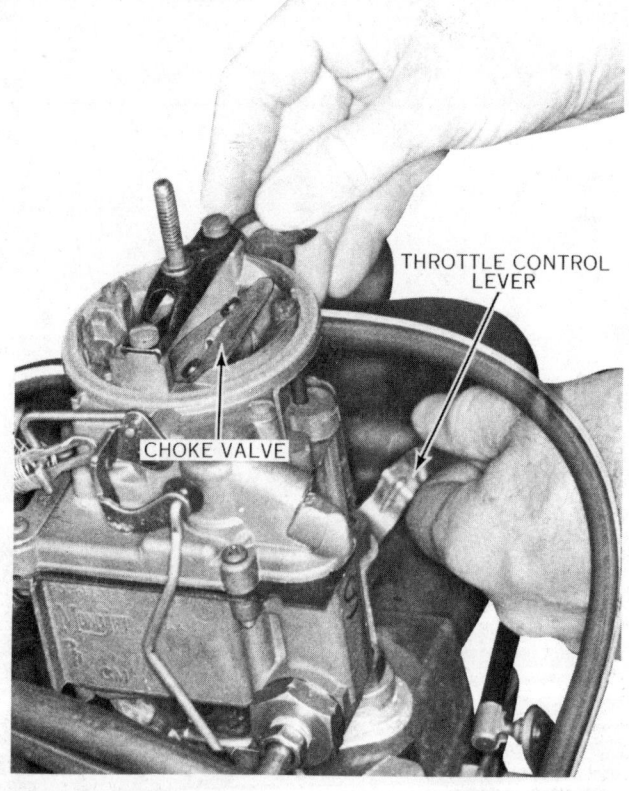

Partially restricting the air flowing into the carburetor with the engine running at about 2,000 rpm is a good way to determine the condition of the air-fuel mixture without elaborate equipment. If the engine speeds up as you partially restrict the air flow, then the mixture is on the lean side. If it slows down, without any speed increase, then the mixture is on the rich side.

CARBURETOR KITS

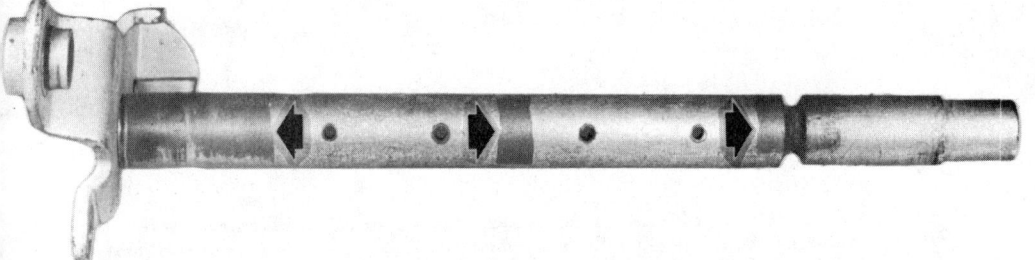

Worn throttle shaft (arrows) shows that all moving parts of the carburetor wear. Cleaning (boiling out) a carburetor and replacing the jets will not correct the problems of wear.

CARBURETOR KITS

Generally, a kit of parts is purchased for a carburetor overhaul, or it can be a "Zip Kit" which contains a minimum of parts. These kits contains all of the jets and gaskets and a set of instructions covering the specifications and bench adjustments for the particular carburetor model you are working on. However, the usefulness of such kits is rather limited in restoring carburetor effectiveness, because most of the parts are not needed and some of them that are

A soot-covered spark plug insulator indicates a rich air-fuel mixture. The combustion chamber temperature was not high enough to burn off the deposits. This carburetor problem is costing you money.

A lean air-fuel mixture will cause higher burning temperatures, and the spark plug will provide this indication.

A metering rod moves up and down each time the throttle is advanced and retarded. As the rod rubs against the main metering jet wall, wear occurs in both parts, and the air-fuel mixture becomes richer. This is the reason why you must replace all wearing parts of the carburetor periodically.

A Holley carburetor uses this diaphragm-type economizer (power) valve. Constant flexing causes it to leak and enrich the idle mixture enough so that you cannot make a lean enough idle mixture adjustment.

CARBURETOR KITS 67

This power valve diaphragm was assembled carelessly, and a vacuum leak occurred where the material was folded over. Take care to do the work right the first time.

The Holley economizer valve contains a diaphragm which eventually leaks due to constant flexing. This enriches the idle mixture to the point where an idle adjustment cannot be made.

needed are not there. For example, most kits contain all of the jets used in the carburetor, but jets never wear and so don't need to be replaced unless drilled out. In fact, it would cause a great deal of damage, if an incorrect jet had been packaged and used in a carburetor. It would be far better to use the old jet than to replace it with an incorrect one.

On the other hand, wearing parts of the carburetor are not always included with kits. To be most effective, a kit of carburetor

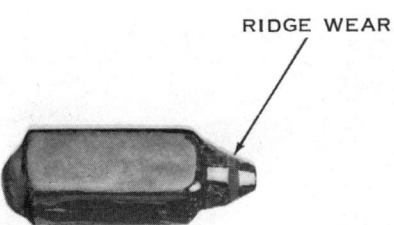

This type of ridge wears into the needle and also into the seat. As mileage accumulates, the ridge gets wider to provide a platform for the dirt particles to rest on, and this is what raises the fuel level in the float bowl, as discussed in the text. This is the reason why it is so important to replace the needle valve-and-seat assembly each time the carburetor is cleaned or overhauled.

Note the bent-over end of this idle mixture adjusting needle, which will prevent you from making an accurate mixture adjustment. It was caused by turning in the needle tightly against the edge of the throttle valve.

parts should always include a set of gaskets and a new needle valve-and-seat assembly, because this is the most wearing part. Also the kit should include all diaphragms, the power jet, and main metering rods and jets because they move which causes wear. Most car-

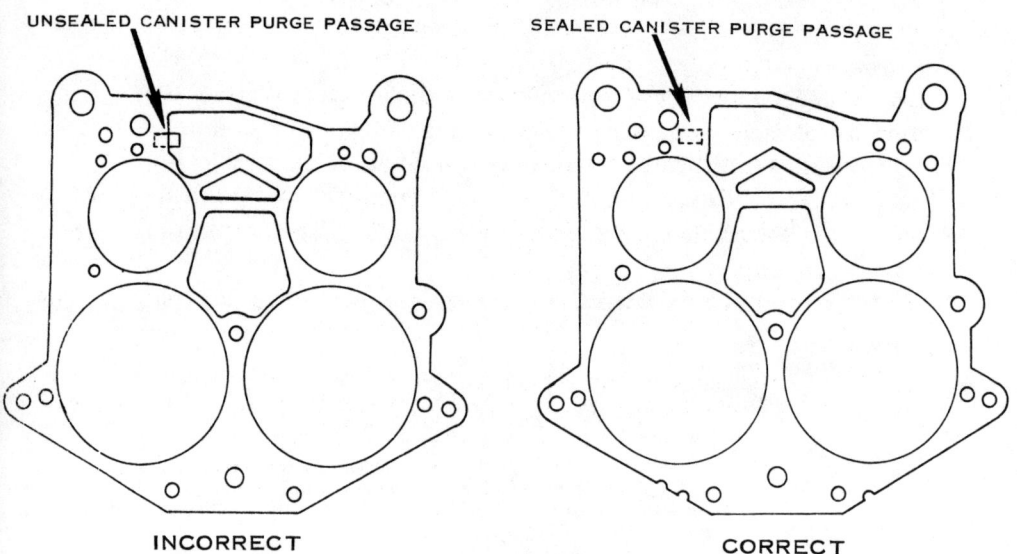

In 1970 Rochester Carburetor Co. changed the gasket used between the float bowl and throttle body of the 4M carburetor due to the addition of a charcoal canister purge port. If the wrong gasket is used on models since 1970, a vacuum leak occurs due to air by-passing the primary throttle valves through the canister purge passageway (left), and the engine idles roughly. This example is used to show that you must compare each gasket carefully with the old one during a carburetor overhaul to be sure that you are installing the correct part and in the correct way.

CARBURETOR CIRCUITS 69

buretors have external linkages which move everytime the accelerator pedal is advanced or retarded, and these wearing parts must be replaced during an overhaul to "time" the operation of the various carburetor circuits precisely. These links are seldom packaged with the kits.

CARBURETOR CIRCUITS

FLOAT SYSTEM

Replace the carburetor fuel inlet filter element every 12,000 miles to minimize the passage of small dust particles past the needle valve-and-seat assembly. Dirt under the needle valve can hold it off the seat and cause flooding, with a resulting loss of gas mileage.

The needle valve-and-seat assembly are a matched set and must always be replaced everytime the carburetor is taken apart; otherwise, leaking will result. This will cause the float level to rise above specifications, resulting in a richer-than-normal air-fuel mixture for all ranges of carburetor operation.

Always adjust the float level to specifications. A higher-than-normal fuel level will cause a rich mixture, just as would a leaking needle valve-and-seat assembly.

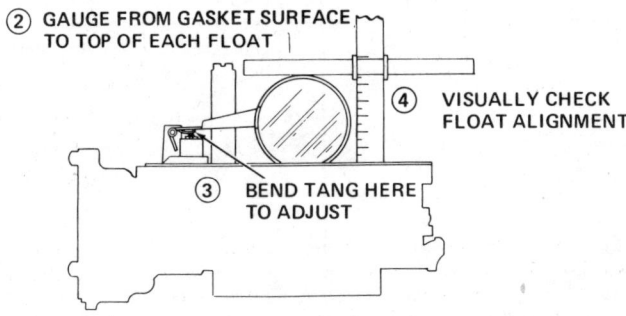

This drawing shows how the float level is measured and adjusted in most carburetors. The specifications will be included in the kit of a parts you buy.

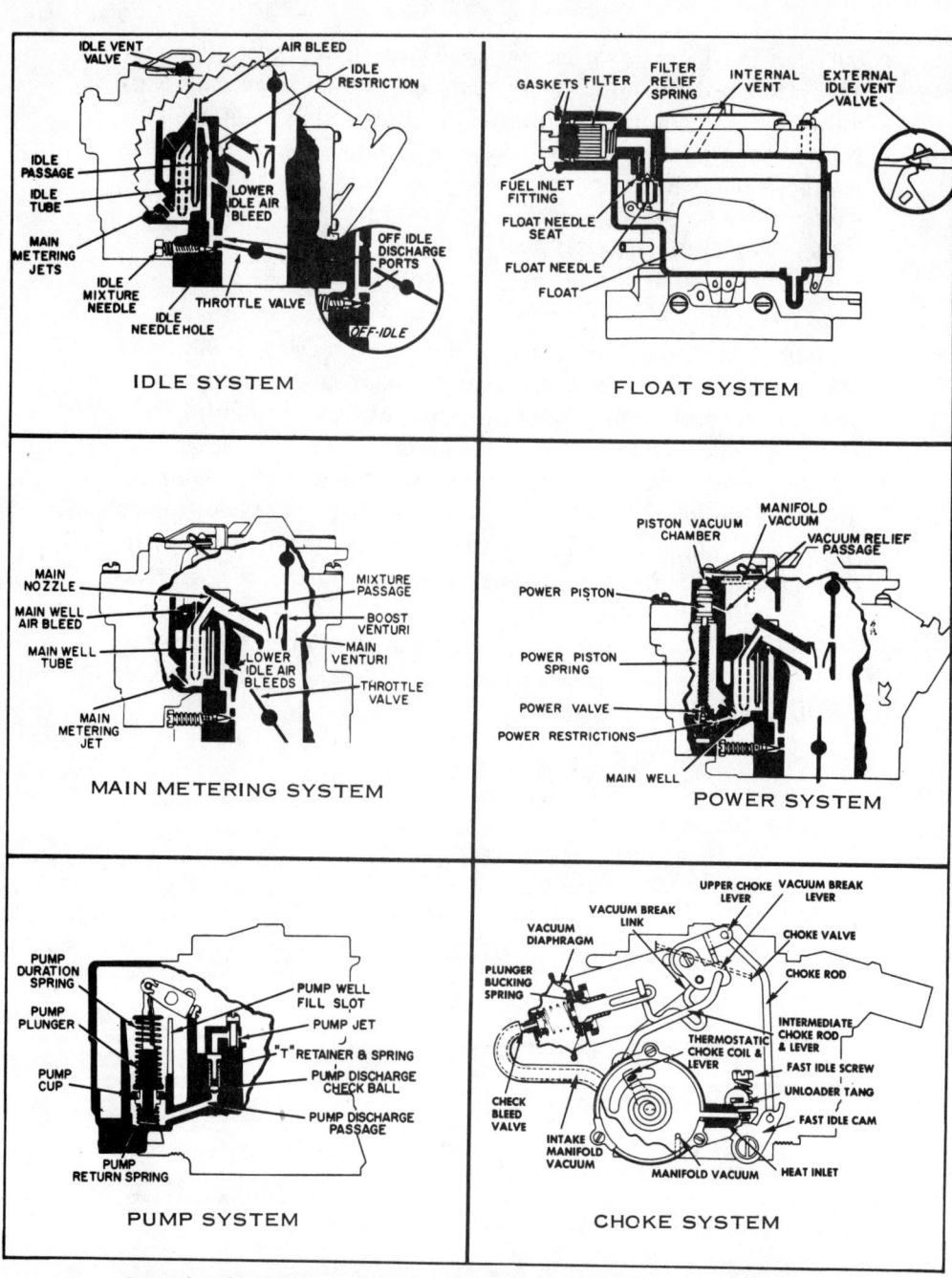

Operating circuits of a Rochester 2GC carburetor. Each of these circuits is discussed in the text so that you can see where the actual part is positioned. Other carburetor models have similar circuits.

CARBURETOR CIRCUITS 71

IDLE SYSTEM

Make sure that the idle mixture adjusting needle is not damaged, or you will not be able to make an accurate adjustment. Because the idle mixture affects the other circuits up through 35 mph, it is essential that the mixture be adjusted accurately, as discussed in the previous section on engine tuning.

If the carburetor has 50,000 miles of operation on it, chances are that the idle air bleed at the top of the carburetor is restricted in size by carbon accumulation. Any reduced air bleed enriches the air-fuel mixture and drops the gas mileage. Cleaning the parts in the carburetor cleaner is the answer to this problem. Or you can replace the air bleed with a new one.

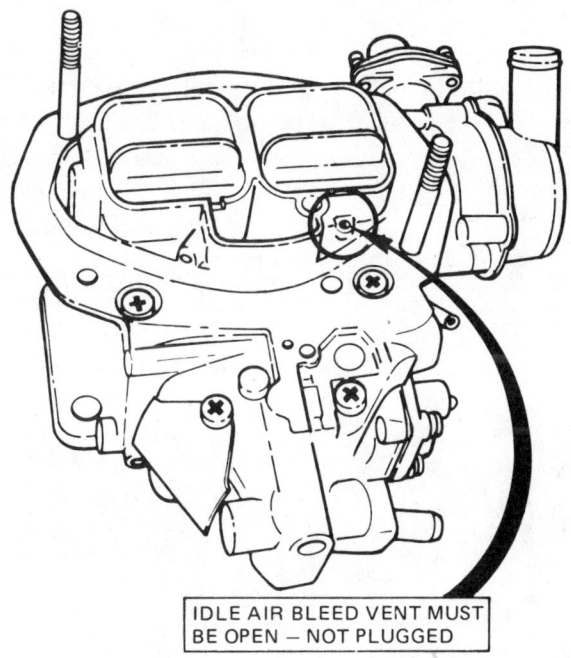

IDLE AIR BLEED VENT MUST BE OPEN – NOT PLUGGED

On some Holley carburetors used on Vega, Pinto, and Mustang, this idle air bleed plugs up and reduces the gas mileage considerably. This is shown as an indication of how small bleed holes in carburetors plug up and must be cleaned to restore operating efficiency and gas mileage.

72   GAS GUZZLER'S GUIDE

Main Metering System

Main metering jets never wear unless they have a metering rod moving up and down in them. In this case, always replace the metering jet and metering rod to restore the air-fuel mixture.

Power System

Sometimes the power jet fails to seat properly, and this causes an enriched mixture which lowers the gas mileage at least 15%. That's wasting $7.50 per month. On some carburetor models, the power jet is vacuum-operated, and this means that it has a diaphragm, which becomes porous from constant flexing. In these cases, it is well worthwhile to replace the power jet whenever you take the carburetor apart.

Some power jet openings are "timed" to throttle opening, and this is accomplished by linkage between the throttle valve and power piston actuating lever. If this linkage wears, then the timing is off. Always replace such worn linkage.

Most power jets are actuated by intake manifold vacuum. In this way, they are sensitive to engine load. When the engine is accelerated hard, the vacuum drops and, when it reaches about 8"Hg, the vacuum piston return spring forces the piston down to

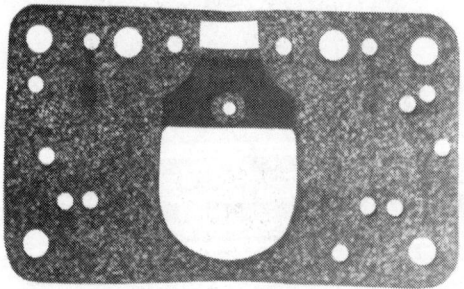
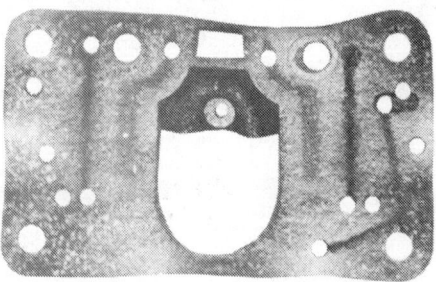

New gasket (left) compared with one removed from the bowl of a Holley carburetor shows why you should use new gaskets when rebuilding a carburetor.

CARBURETOR CIRCUITS 73

open the power jet. In a number of cases, the incorrect carburetor flange gasket was furnished in the kit of parts. If the gasket blocks the vacuum source to the power jet, then the mixture will be enriched throughout the driving range, resulting in a 15% reduction in economy, for a cost of $90.00 per year. Take the time to trace out this vital passageway to make sure that the flange gasket has the proper hole and that you're installing it so that the hold matches the vacuum passageways in the intake manifold flange and in the carburetor casting.

ACCELERATING SYSTEM

The accelerator pump forces a stream of gasoline into the air stream everytime the throttle is advanced to accelerate. Dirt can cause the two check valves to stick, and this will result in a "flat spot" during acceleration, which affects performance to a considerable degree. Clean or replace the check valves at each overhaul.

The pump leather dries out after long use in gasoline, therefore, the plunger should be replaced everytime the carburetor is taken apart for service. Also, the linkage connecting the accelerator pump plunger with the throttle valve wears, and this linkage must be replaced at the same time. When making the bench adjustments, the

The accelerator pump leather wears through like this in some models due to the constant movement of the pump plunger everytime the throttle is moved. Naturally, such worn parts must be replaced.

position of the piston is always measured with regard to the throttle, and this measurement affects the stroke of the piston and, therefore, the performance of the engine.

Some carburetors have provision for changing the piston stroke length for summer and winter driving. If you are interested in saving money, especially in city driving, and are willing to sacrifice some performance in the winter months, you can move the pump stroke adjustment to the shortest one. Make a few tests to determine whether the fuel savings is worthwhile.

AUTOMATIC CHOKE

The thermostatic coil within the choke is heated by the exhaust

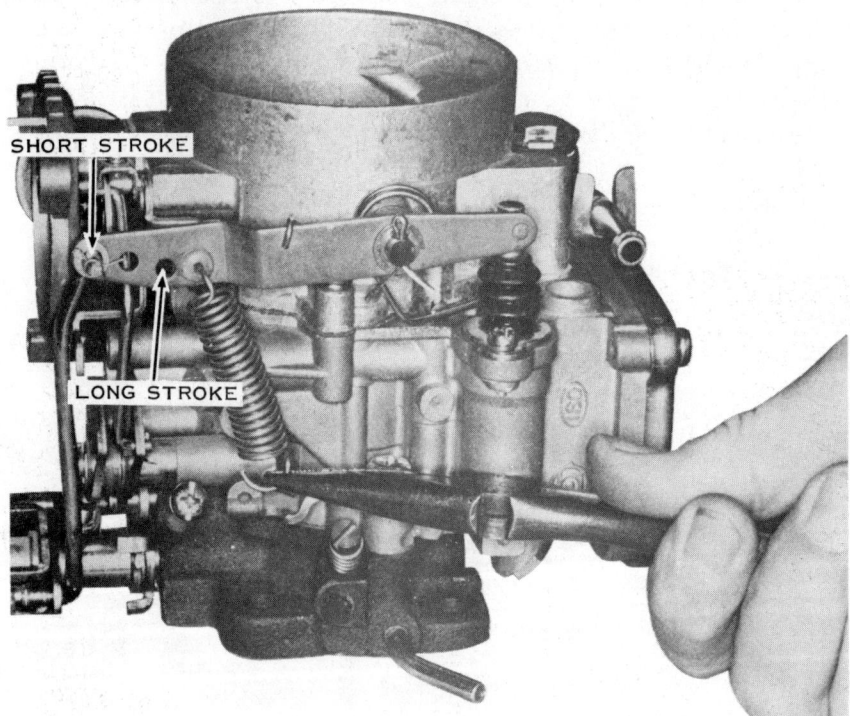

Some carburetors have stroke-length adjustment holes, and it is possible to shorten the stroke by moving the operating rod to another hole in the lever; this can increase your city gas mileage considerably, as discussed in the text.

CARBURETOR CIRCUITS 75

Corroded automatic choke thermostatic coil, evidence of a defective heat tube leaking exhaust gas into the choke mechanism. You just can't get any mileage when the choke sticks closed most of the time.

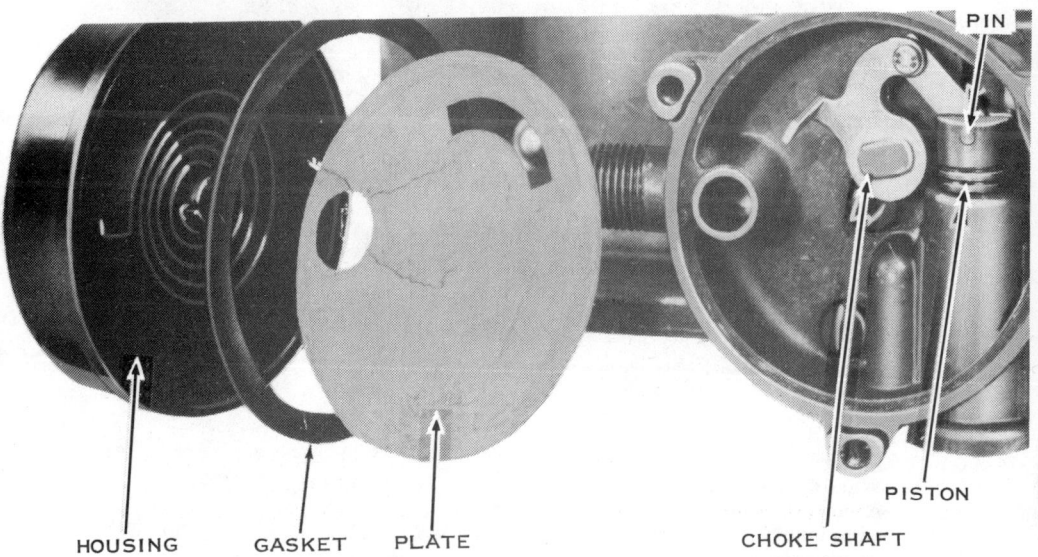

An automatic choke can give gas mileage problems when it becomes sluggish in action. This exploded view shows the parts of a conventional automatic choke. The cover is held to the air horn of the carburetor by three washers and screws. The adjustment mentioned in the text can be made by loosening the cover screws and rotating the cover one notch to the lean side for increased gas mileage if you make a lot of cold starts.

and this can cause the formation of gum, which will result in a sticking choke. Depending on how bad the situation is, you can be losing 100-200 dollars per year in this sticky situation. Cleaning the choke mechanism is the only solution to restore choke performance.

The setting of the automatic choke is always made to ensure good driveability with a cold engine. It is possible to save some bucks if you are willing to sacrifice some of the warm-up driveability, especially in the winter months. Try loosening the choke cover screws and turning the choke cover one notch to the lean setting. You may like the savings.

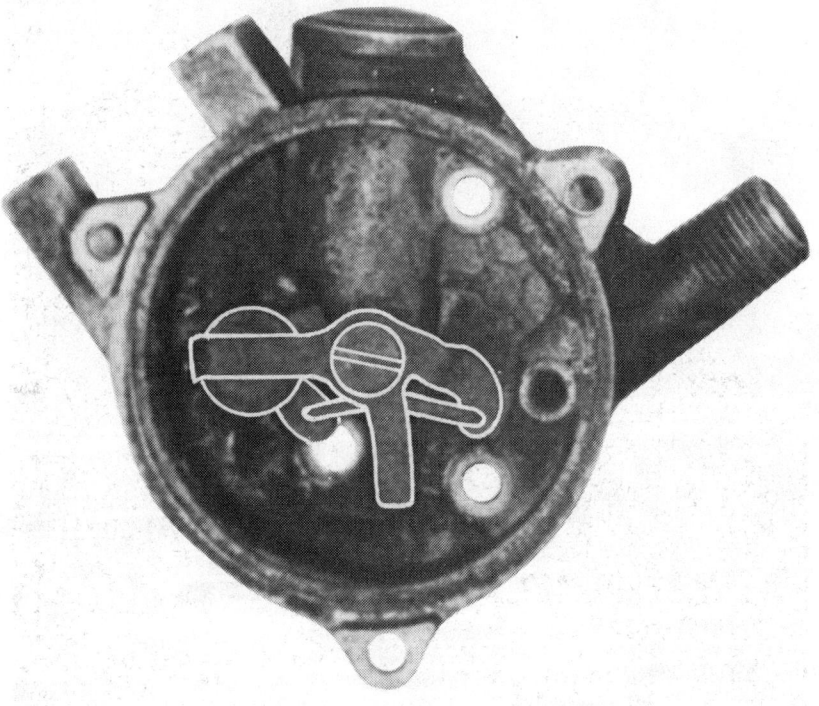

When the inside of the choke housing is carbonized up like this one, the operating mechanism is stuck and improper choke action will occur. Such a condition is caused by a leaking heat intake pipe; exhaust fumes are getting into the choke housing to cause these deposits. Gas mileage on this engine will be very low.

CARBURETOR ADJUSTMENTS

The only adjustments on modern carburetors are those that can be made at idle speed. However, carburetor design is such that the effects of these adjustments have a definite influence on the air-fuel mixture delivered to the engine up to 35 mph, and this can affect your gas mileage to a considerable degree, especially if you drive in the city a great deal.

A vacuum gauge is handy for adjusting the idle mixture screws. Make the adjustment to obtain the highest vacuum reading. CAUTION: Most manufacturers require that you have the air cleaner in place during this adjustment.

Making the idle speed adjustment on a Vega carburetor.

CARBURETOR ADJUSTMENTS 79

EMISSION-CONTROLLED ENGINES

All emission testing by highway patrols or official smog stations is done at idle speed so it is important that you follow these instructions precisely if you don't want to run afoul of the law. Really, it's no big deal to get your emissions within legal limits, **without expensive tune-up equipment**. All you need is our familiar and inexpensive vacuum gauge and a tachometer to adjust the carburetor mixture using the "lean-drop method" described below.

Lean-Drop Method

First check the tuning decal for the idle speed specifications, which is always given as two figures, such as 650-750. This means that you have 100 rpm leeway in making the idle speed adjustment. The engine can be running between 650 and 750 rpm and still be within the specified limits. Hook up your vacuum gauge to a tap on the intake manifold and the tachometer to the primary distributor

GM 101-1 140 Cu. In. Federal	Transmission	
	Automatic	Manual
Exhaust Emission Control System Timing (^{O}BTC @ RPM) Lean Drop Idle Mixture (RPM)	CCS–EGR 12° @ 750 800-750	CCS–EGR 10° @ 700 800-700
GM 101-1 140 Cu. In. NB–2		
Exhaust Emission Control System Timing (^{O}BTC @ RPM) Lean Drop Idle Mixture (RPM)	CCS–EGR 8° @ 750 800-750	CCS–EGR 8° @ 700 800-700
GM 101-2 140 Cu. In. – Nationwide –		
Exhaust Emission Control System Timing (^{O}BTC @ RPM) Lean Drop Idle Mixture (RPM)	CCS–EGR 12° @ 750 800-750	CCS–EGR 10° @ 700 800-700

Examples of three emission-control decals. Note that the Lean-Drop Idle Mixture specification is a spread of two speeds, as discussed in the text.

terminal of the ignition coil. Now start the engine and warm it to operating temperature. Disconnect the hose leading to the evaporative emission-control system charcoal canister to avoid having gasoline-laden vapors entering the induction system and affecting your carburetor adjustments. Or you can remove the gas tank cap to avoid pressurizing the system which will minimize the flow of vapors from the canister.

Turn the idle speed adjusting (throttle stop) screw to obtain the higher of the speeds shown on the tuning decal which, in the case above, is 750 rpm. Now adjust the idle mixture screw to obtain the highest vacuum reading (or the best running engine if you are not

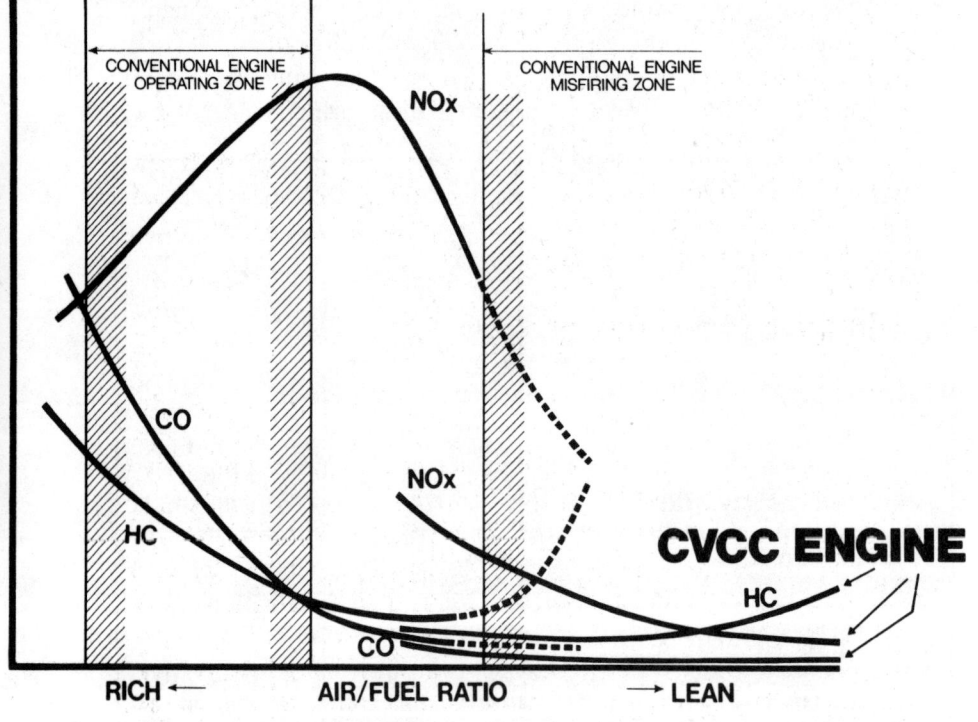

This graph shows the effect of changing the idle mixture on the three exhaust emissions. Note that NOx is increased as you lean the mixture of a conventional engine. Also shown is the effect of changing the mixture of a Honda CVCC engine, which is extremely clean as regards to exhaust emissions.

using a vacuum gauge). With a two- or four-barrel carburetor, you must adjust both screws the same amount. *NOTE: If you are tuning a modern emission-controlled engine, it will be necessary to break off the tab on the plastic cap covering the mixture adjusting screw to make the adjustment.* If engine speed changes from our initial setting of 750 rpm, it is essential to adjust the idle speed screw again to return engine speed to the previous figure of 750 rpm.

To make the final mixture and speed adjustments so your emissions are within legal limits, turn the idle mixture adjusting screw(s) clockwise (lean) to lower your idle speed to the lower of the two figures specified on the tuning decal, which in this case would be 650 rpm for a 100-rpm reduction. **CAUTION: Don't touch the idle speed screw.** Make this engine speed reduction by leaning the mixture with the mixture adjusting screw to keep your hydrocarbon emissions within legal limits.

Replace the plastic limit caps with new ones that are sold for service so that the job is done legally. In all cases, the manufacturer supplies the idle speed specifications with the proper spread so that the "lean-drop method" can be used to obtain legal emission limits without an expensive CO meter.

EMISSION-CONTROL SYSTEMS

Crankcase, exhaust, and evaporative emission-control systems are an inherent part of modern engines, and some of these systems can be gas gulpers, especially if they are not functioning properly. This section will be for the mechanically handy person who wants to check out the functioning of the systems on his car to isolate a recent gas mileage drop to determine whether the trouble is due to a malfunction in an emission-control system, or to some ignition/fuel system problems.

In all cases, some quick tests are provided so that you can check the efficiency of the system. If you find that your engine does not pass this test, then you should return the vehicle to your dealership for more detailed testing to determine the exact part that has failed. This will restore your gas mileage as well as make your vehicle conform to legal emission limits.

CRANKCASE EMISSION-CONTROL SYSTEMS

All of these systems depend on an air bleed in the intake manifold connected to a sealed-type crankcase. Its purpose is to draw the blow-by gases from the crankcase into the intake manifold so that they can be consumed in the combustion process. Some systems use a calibrated bleed hole, but most of them have a PCV (Positive Crankcase Ventilation) valve in the circuit.

If the PCV valve or the calibrated bleed hole in the intake manifold clogs, then the designed amount of air will be shut off from the intake manifold, and your gas mileage will decrease at least 10%; this is going to cost you about $5.00 per month in added fuel costs.

System Quick Test

With the engine running at about 1,000 rpm, use a pair of pliers

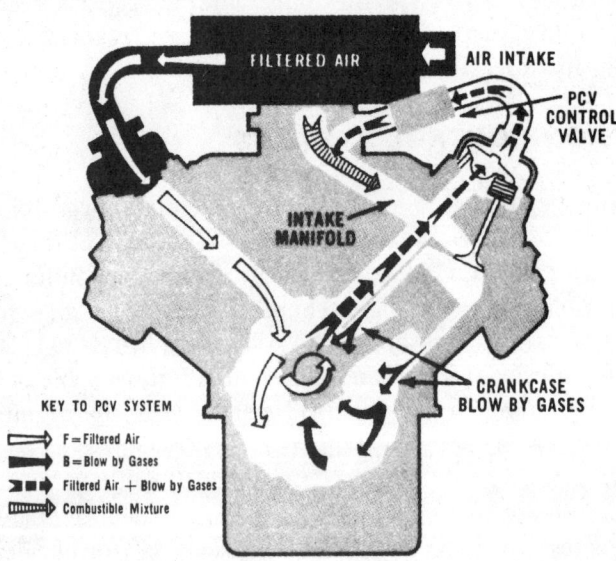

This diagram shows how the air and blow-by gases from the crankcase pass through the PCV valve and into the combustion chamber for burning.

CRANKCASE EMISSION-CONTROL SYSTEM 83

to clamp off the hose running from the PCV valve to the intake manifold. Engine speed should drop about 60 rpm if the system is functioning properly.

Another quick test is to remove the oil filler cap with the engine idling; engine speed should increase at least 60 rpm to indicate the fact that you are allowing extra air to enter the system and relieve some of the vacuum designed into it. The added air flow leans the air-fuel mixture, and the engine should run slightly faster. If the **System Quick Test** indicates trouble, then make the detailed tests which follow to isolate it.

A quick test of the efficiency of the PCV system is to clamp off the hose running to the PCV valve. Engine speed should drop about 60 rpm if the system is functioning properly.

Service Procedures

① Remove the PCV valve from the rocker arm cover and shake it. You should hear a clicking noise if the valve is free. If not, replace the valve. **CAUTION: Don't attempt to clean the valve. Also, make sure that you replace it with one having the correct part number, as each has a different calibration.**

② Start the engine and a hissing noise should be heard as air passes through the valve. A strong vacuum should be felt when you place your finger over the valve inlet.

CRANKCASE EMISSION-CONTROL SYSTEM

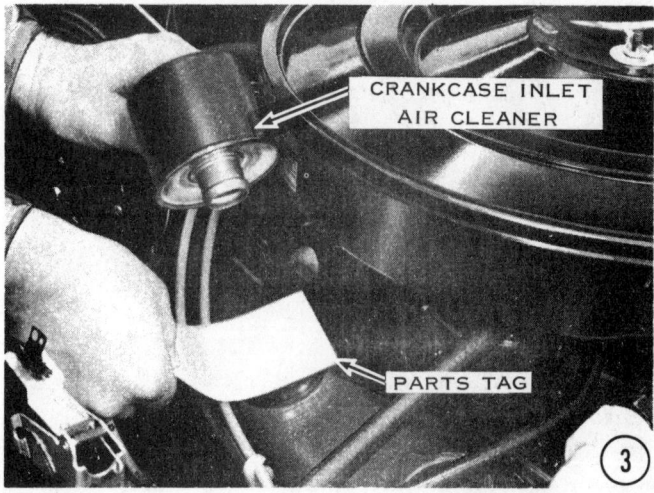

③ Reinstall the valve, and then remove the oil filler cap. Start the engine and hold a piece of cardboard over the opening in the rocker arm cover. If the system is functioning properly, the cardboard will be sucked against the opening with a noticeable force. If the cardboard is not sucked against the opening with a new PCV valve, it is necessary to clean the hoses, vent tube, and passageway in the lower part of the carburetor.

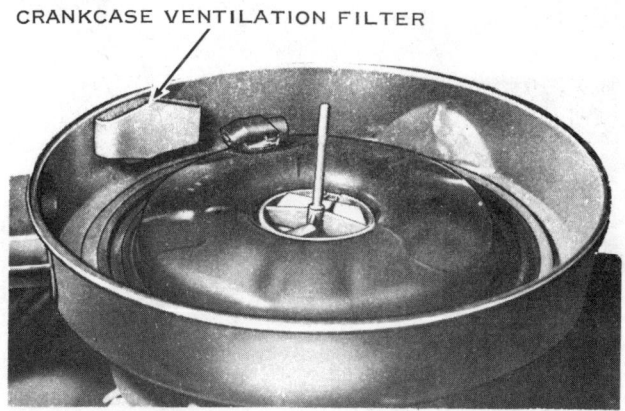

If the crankcase ventilation system has a filter element in the air cleaner shell, it must be cleaned or replaced every 12,000 miles.

EXHAUST EMISSION-CONTROL SYSTEMS

There are three basic systems: (1) Transmission-Controlled Spark (TCS), (2) Spark-Delay System (SDS), and (3) Exhaust-Gas Recirculation (EGR). There are other exhaust emission-control systems (and evaporative emission-control systems), but they have more effect on emissions than mileage; therefore, we will not discuss them in this book.

TRANSMISSION-CONTROLLED SPARK

The TCS system, also known as the NOx system on Chrysler engines, Transmission-Regulated Spark (TRS) on Ford products, or Speed-Controlled Spark (SCS) on others, is designed to deny vacuum to the distributor vacuum-advance actuator during certain engine-operating modes. Generally, vacuum is denied until the transmission is shifted into high gear (TCS) or until it reaches a designed vehicle speed of about 35 mph (SCS). If this system is not functioning properly, you can be losing as much as 25° vacuum advance, and this can be costing you about 40% efficiency in average-

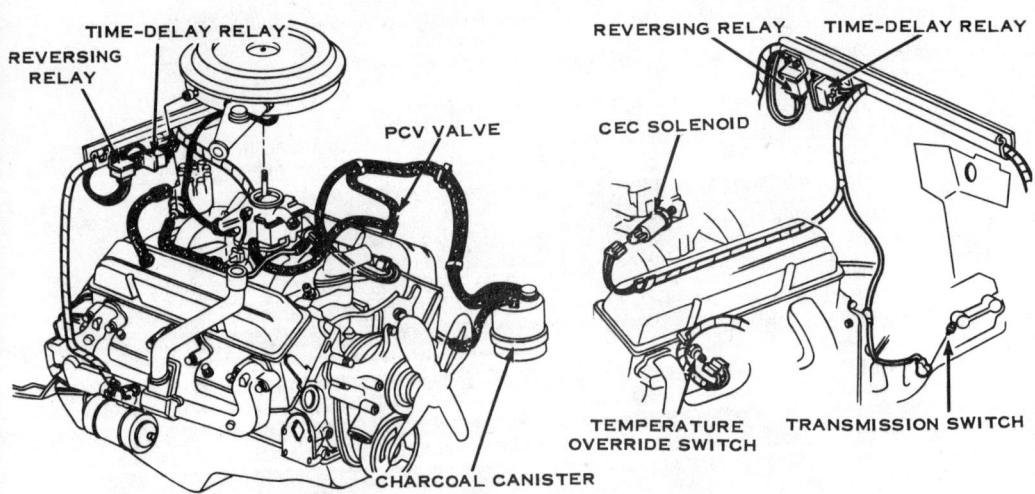

Layout of the various control units making up the TCS system used on the 1971 Chevrolet V-8 engine. Other models have the parts placed in similar positions.

TRANSMISSION-CONTROLLED SPARK SYSTEM 87

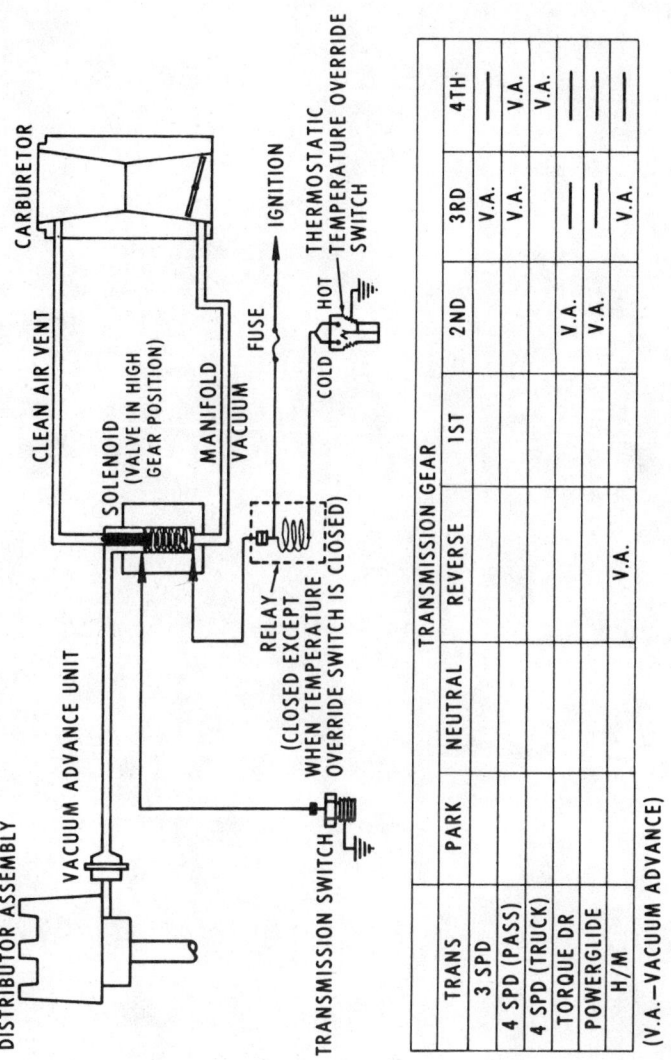

Circuit diagram of the Transmission-Controlled Spark system used on a 1970 Chevrolet V-8 engine. This same system is used on most G.M. cars, as discussed in the text. If there should be a malfunction in the system, you may not be getting vacuum to the distributor vacuum-advance diaphragm, and this would lower your gas mileage considerably during normal cruising.

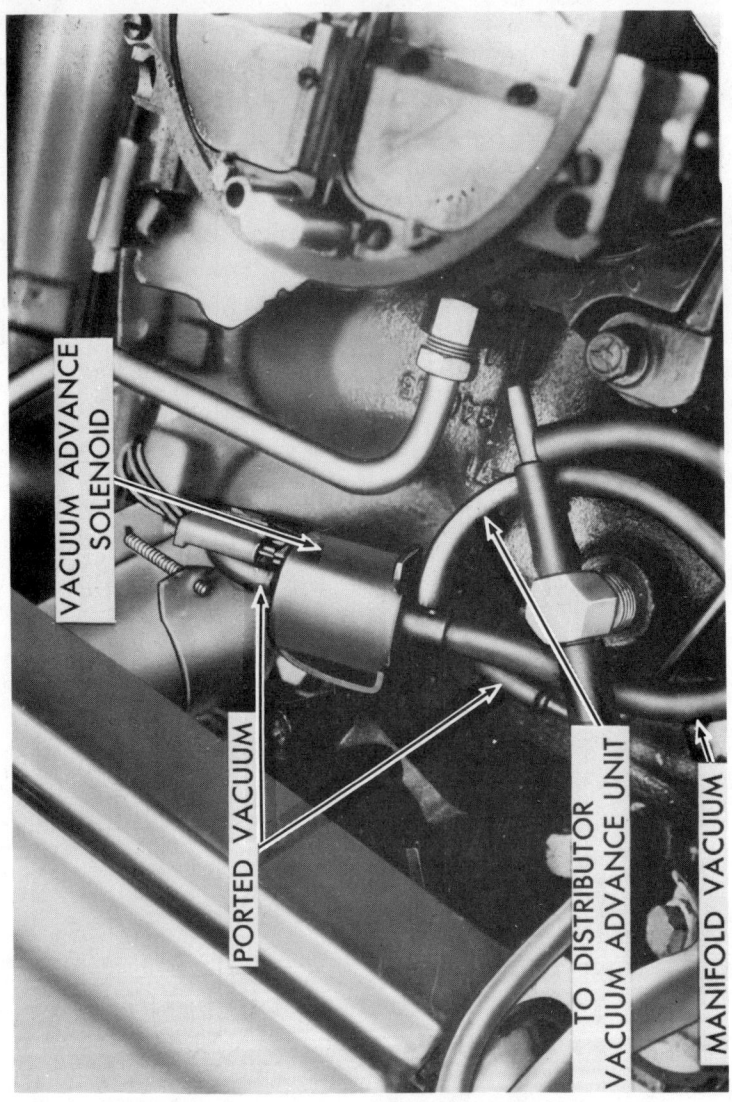

Typical TCS solenoid placement and position of the hoses that supply vacuum to the distributor. In many cases, it is only necessary to pull the connector from the TCS solenoid to obtain vacuum for testing purposes, as discussed in the text.

TRANSMISSION-CONTROLLED SPARK SYSTEM 89

speed, high-gear operation, where the TCS system should not function. A defective TCS system can cost you almost $250.00 per year in added fuel costs. How can you check out this system? Read on!

System Basic Tests

Disconnect the vacuum hose leading to the vacuum-advance actuator on the distributor and connect it to a vacuum gauge that can be viewed from the driver's seat. (This needs a long hose.) If you already have the suggested vacuum driving gauge mounted on the dash (See Chapter 4), it is relatively easy to connect a jumper hose from the distributor hose to the existing hose leading to your dash-mounted gauge.

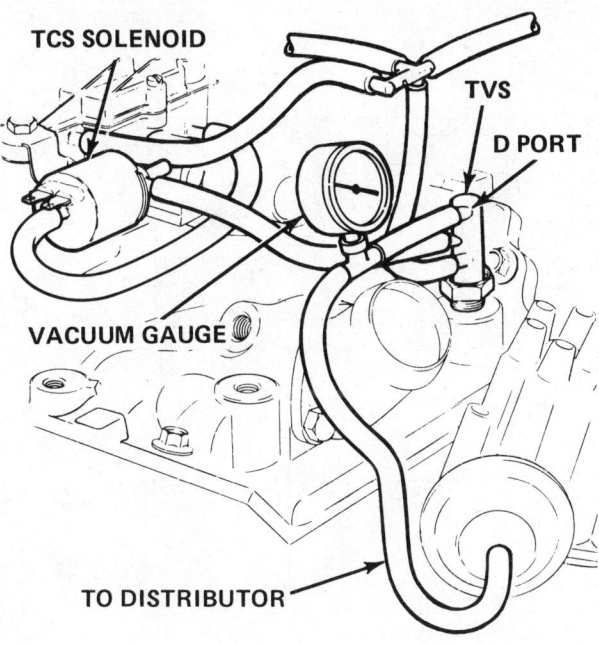

To check the TCS system, hook a vacuum gauge into the hose leading to the distributor vacuum-advance unit. You should obtain vacuum in high gear with a manual transmission or in reverse with an automatic. If you don't get vacuum, then you're running on retarded timing, and this is costing you lost engine performance and poor gas mileage.

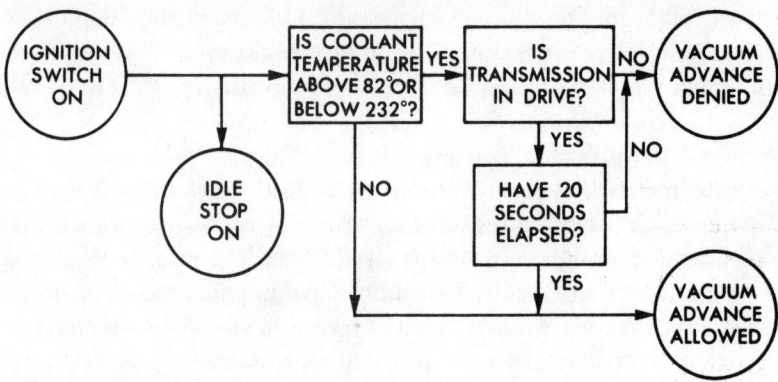

Vacuum diagram to show the conditions under which vacuum is allowed or denied to the distributor vacuum-advance unit.

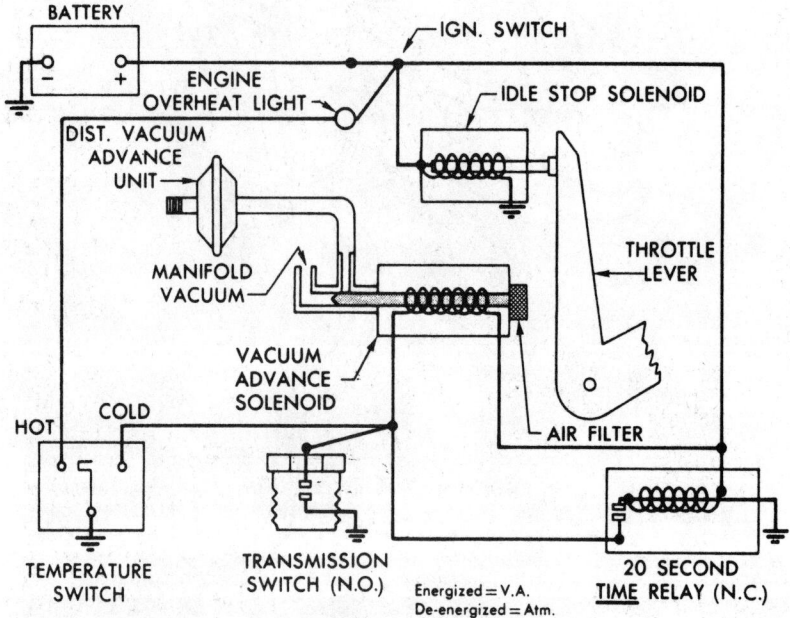

Circuit diagram of the TCS system used on a 1973 Chevrolet V-8 engine in the low-gear operating mode. Note that the temperature switch, transmission switch, and time-delay relay are open circuited to deprive the distributor of vacuum in the low-gear operating mode.

TRANSMISSION-CONTROLLED SPARK SYSTEM 91

Now drive the vehicle until it is thoroughly warmed. With the ambient air temperature above 68°F, and the engine running over 1,500 rpm, you should get full vacuum to the gauge only after shifting into HIGH or DRIVE (or get above 35 mph with an SCS system). **CAUTION: You may have to wait for almost 60 seconds for full vacuum to appear on some of the later engines with a time-delay relay.** The vacuum gauge must drop to zero when you shift into NEUTRAL. The same test can be made in the garage by jacking up the rear axle and supporting the car on safety stands during the test. **CAUTION: Make sure the parking brake is firmly applied and the front wheels blocked for safety reasons.**

If the system passes these two checks, no further testing is needed.

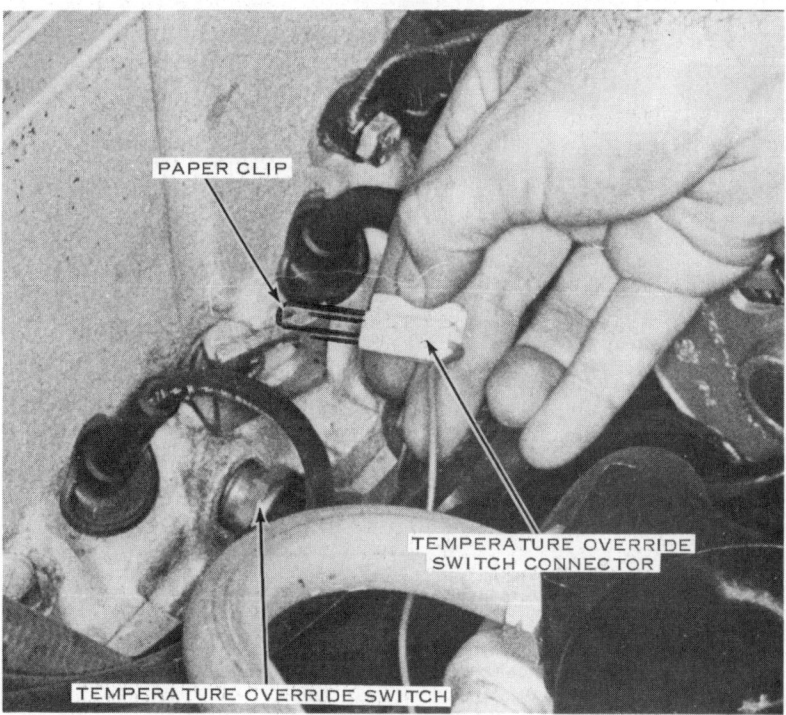

Grounding the Vega temperature override switch wires restores vacuum to the distributor vacuum-advance unit for testing purposes.

GAS GUZZLER'S GUIDE

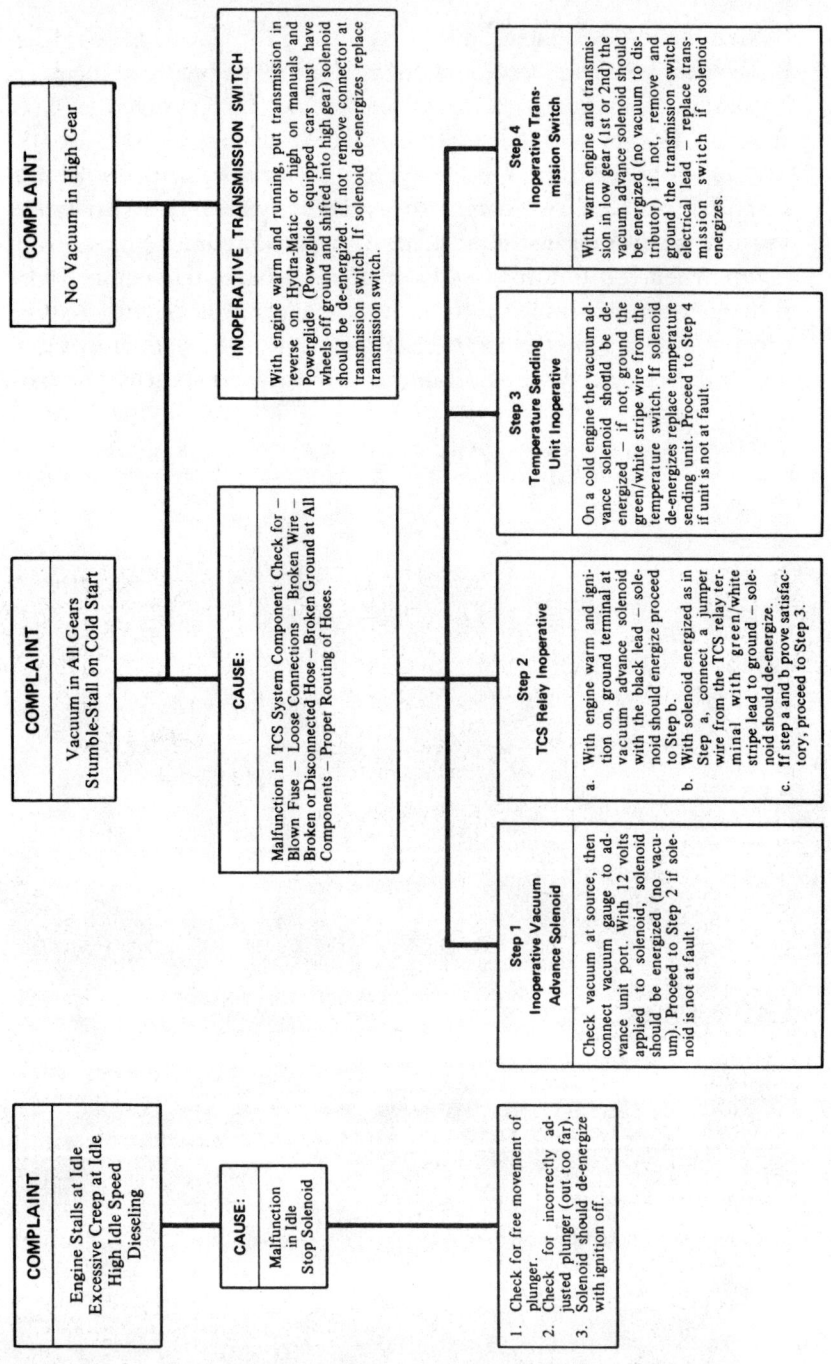

Transmission-Controlled Spark (TCS) system troubleshooting guide.

TRANSMISSION-CONTROLLED SPARK SYSTEM 93

System Isolation Test

To isolate trouble between the ignition/fuel systems and the TCS system, it is often desirable to disconnect the TCS system and test-drive the vehicle without it in operation to determine whether the problem is resolved. **CAUTION: After making the following tests, be sure to reconnect the system properly so that the vehicle conforms to legal emission standards.**

On most early G.M. TCS systems, the solenoid was energized to provide vacuum advance and, on later models, it is de-energized to provide vacuum advance; therefore, it is necessary to determine which system you are working on for the proper procedure.

You can pull off the TCS solenoid connector on many General Motors engines to restore vacuum to the distributor for testing purposes.

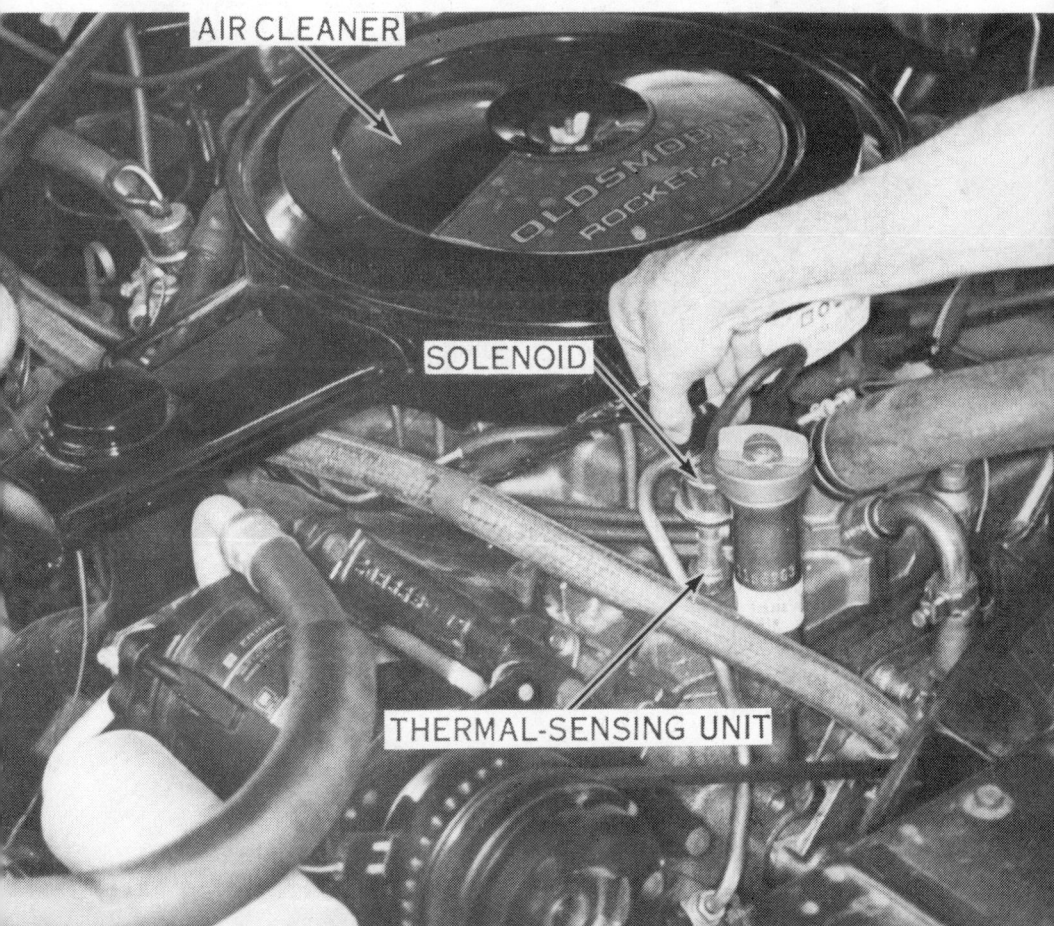

94 GAS GUZZLER'S GUIDE

All Ford, Chrysler, and American Motors systems provide vacuum advance when the solenoid is de-energized. On these systems, it is only necessary to disconnect the feed wire to the solenoid and drive the vehicle without the system in operation. This also works on all 1970-71 General Motors cars, except Chevrolet, in which case, it is easiest to disconnect the wires at the heat sensor in the cylinder head and ground them with a jumper wire to provide full vacuum advance for testing purposes. After testing, reconnect the wires to restore the TCS system's functions for reducing exhaust emissions.

On engines where the solenoid must be energized to provide full

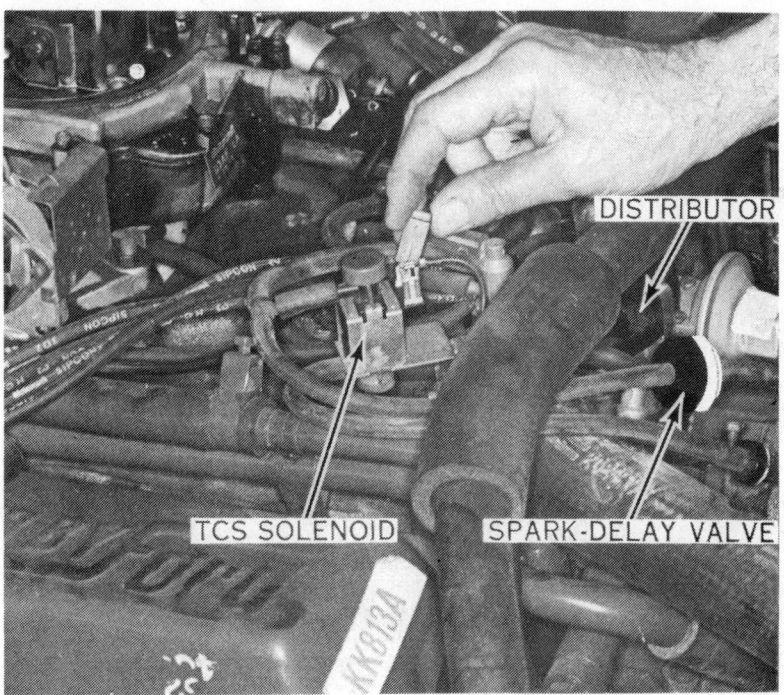

On Ford engines, you can disconnect either wire leading to the TCS solenoid to disarm the system for testing purposes.

TRANSMISSION-CONTROLLED SPARK SYSTEM

vacuum advance for testing purposes, it is first necessary to determine which of the two wires to the solenoid is the feed wire and which is the grounding one; otherwise, you could ground the wrong wire and blow a fuse to complicate the testing procedure. Check both wires to the solenoid with a test lamp or voltmeter to determine which one is the feed wire, and then use a jumper wire to ground the other solenoid terminal to provide full vacuum advance for testing purposes. After testing, disconnect the jumper wire and reconnect the feed wire to the TCS solenoid to restore the TCS system's functions for reducing exhaust emissions.

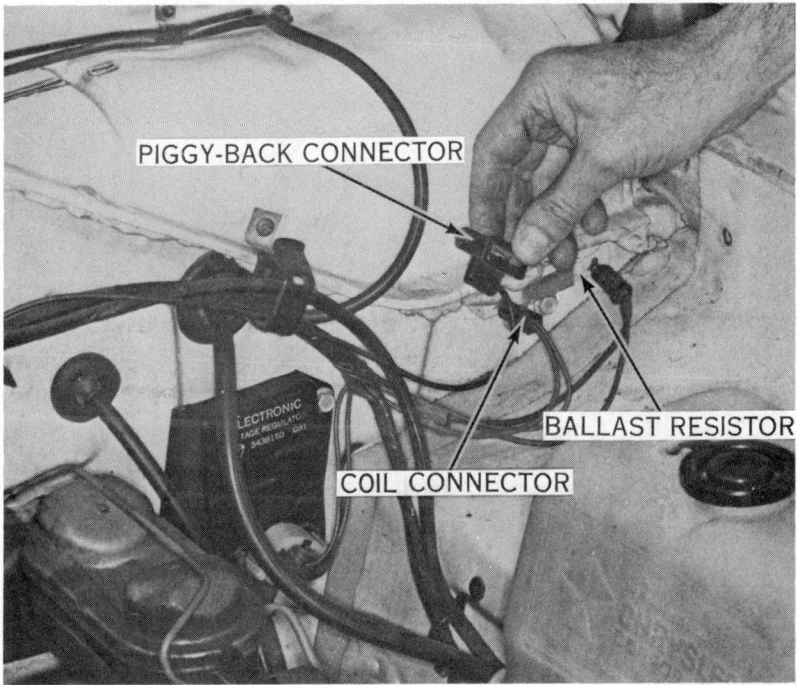

The Chrysler NOx system can be disarmed for testing purposes by disconnecting the piggyback connector on the ballast resistor, and then reconnecting the wires to the terminal.

The Spark-Delay Valve (SDV) should always have the black side connected to the source of vacuum. This sintered metal valve must be replaced every 10,000 miles, or engine performance and gas mileage will be adversely affected.

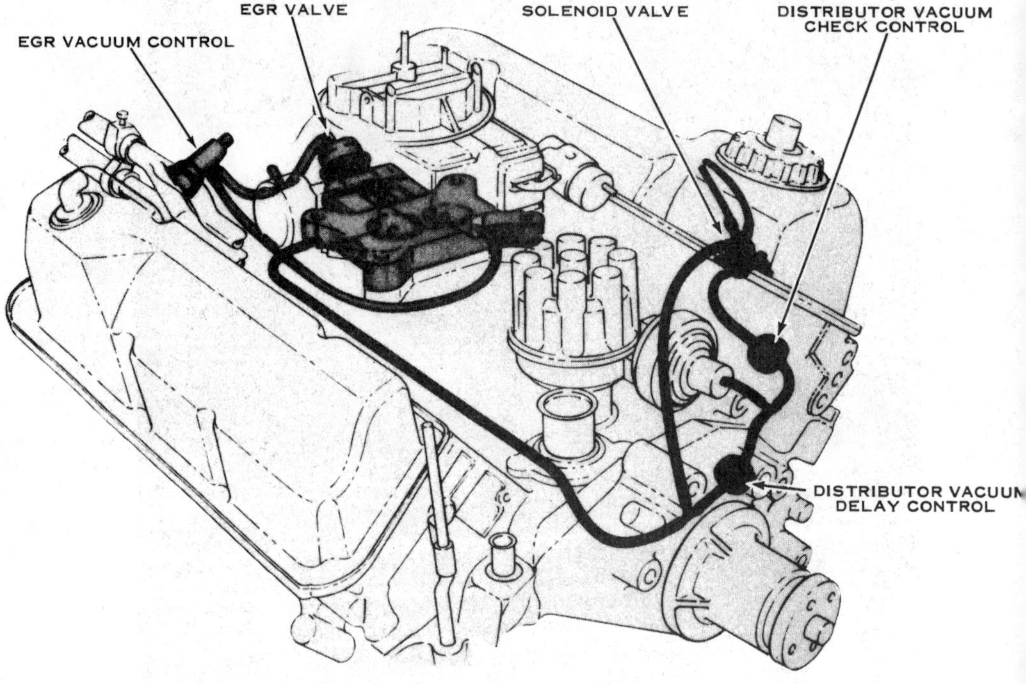

Typical Spark-Delay (SD) system as used on a 1973 Ford V-8 engine. Note that this system is used in conjunction with an EGR valve, which is controlled by a Thermal Vacuum-Switching (TVS) valve. EGR is limited to operating conditions when the engine has reached normal temperatures.

SERVICING THE SPARK-DELAY SYSTEM

SPARK-DELAY SYSTEMS (SDS)

All of these systems delay vacuum to the distributor vacuum-advance actuator for a designed period of time, which can be as long as 60 seconds. Generally, these valves are made of sintered bronze discs, with very small pores for vacuum to bleed through. The more discs stacked inside of the SDS valve, the longer it takes for vacuum to pass through it.

Chrysler calls its delay system Orifice Spark Advance Control (OSAC), and it functions exactly like the SDS.

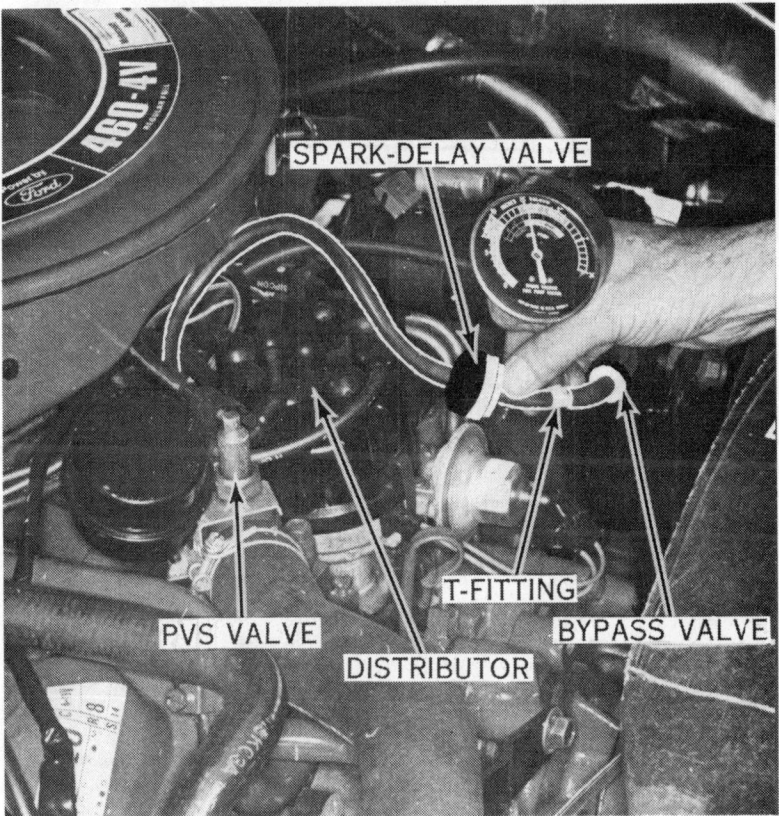

The Ford Spark-Delay System (SDS) can be tested with a vacuum gauge, as discussed in the text.

System Basic Test

Disconnect the vacuum hose leading to the vacuum-advance actuator on the distributor and connect it to a vacuum gauge. Start the engine and accelerate it to about 1,500 rpm. The vacuum gauge should show no reading for a few seconds and then it should build up slowly to over 15"Hg, **provided you keep the throttle steady**. Everytime you accelerate the engine, the vacuum reading will drop and it will take time to reach the normal level again **if you hold the throttle steady**.

Testing the Spark-Delay valve on a STP Retro-fit installation, using a vacuum gauge as discussed in the text.

SERVICING THE SPARK-DELAY SYSTEM

SERVICE PROCEDURES

These valves are subject to blockage by particles of dust that build up as air, influenced by throttle action, continuously moves both ways through the valve. After about 10,000 miles, the valve will be obstructed enough to reduce your vacuum advance, and this can lower your gas mileage as much as 40% for a loss of about 35 gallons of gasoline per month. Replacing the SDS valve every 10,000 miles is good insurance against this happening.

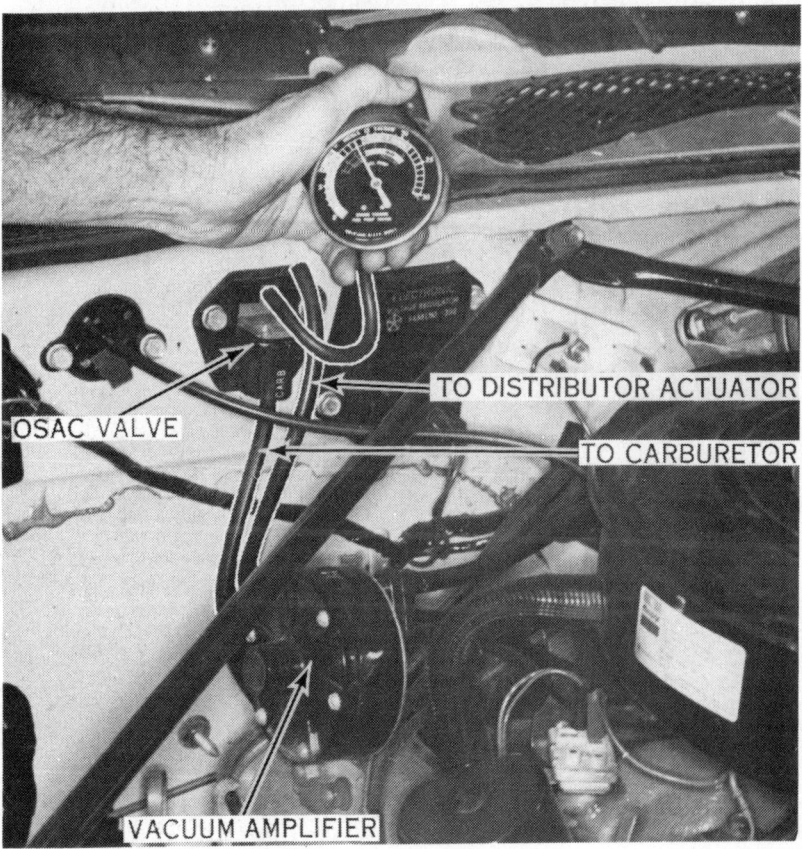

Testing the Chrysler Orifice Spark Advance Control (OSAC) valve with a vacuum gauge. The test is similar to that for the SDS valve discussed in the text.

EXHAUST-GAS RECIRCULATION (EGR)

The purpose of the EGR system is to control the emissions of oxides of nitrogen (NOx). It does this by returning a small amount of exhaust gas into the intake manifold to reduce the combustion chamber temperature below a critical level so that less NOx is formed.

Because the introduction of exhaust gas reduces combustion efficiency, the system is designed to be inoperative during cold engine starts and low-speed operation in order to restore driveability. It introduces the maximum amount of exhaust gas when the engine is heavily loaded, as during hard acceleration when combustion chamber temperatures reach higher-than-normal levels.

If the control system is defective and allows EGR while the

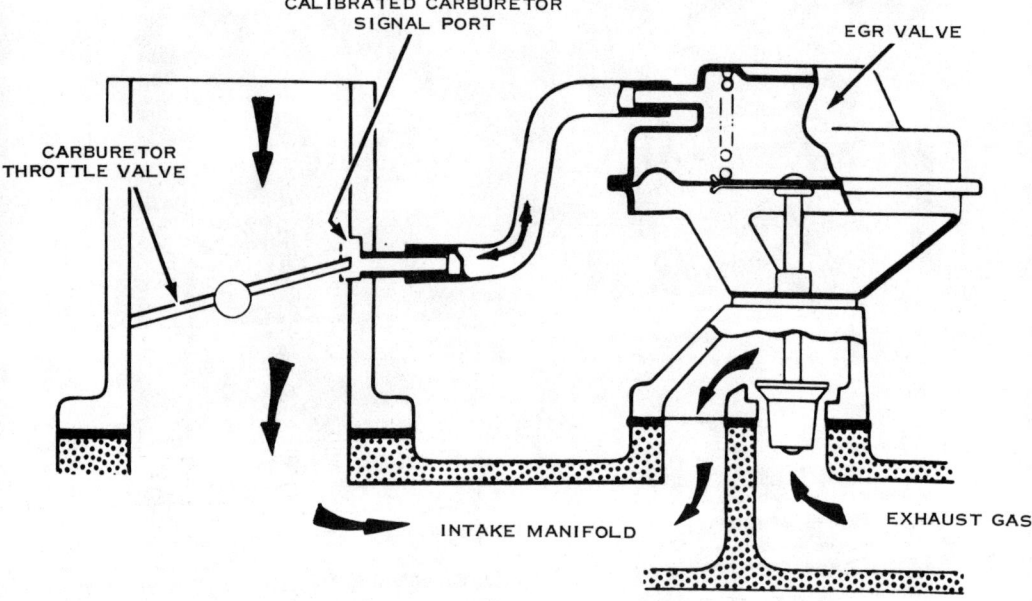

The EGR system allows calibrated amounts of exhaust gas to enter the intake manifold for reducing the temperature of combustion to minimize the formation of NOx.

SERVICING THE SPARK-DELAY SYSTEM 101

engine is cold, it will be difficult to start. If the EGR valve leaks, idling quality will be seriously affected. In any case, a defective EGR valve or control system will lower gas mileage accordingly.

System Quick Tests

To test the EGR valve, open the throttle of a thoroughly warmed engine and observe the EGR valve shaft, which should move up and down as the throttle is moved back and forth. If it doesn't, the EGR valve or the control system is defective.

To test the EGR control system, start the engine, clamp off the vacuum supply hose to the EGR valve, and then raise engine speed to about 1,500 rpm. Release the clamping hose pressure, and engine speed should drop about 150 rpm to reflect the passage of exhaust gas into the intake manifold as the EGR system becomes effective. If it doesn't, the control system is defective.

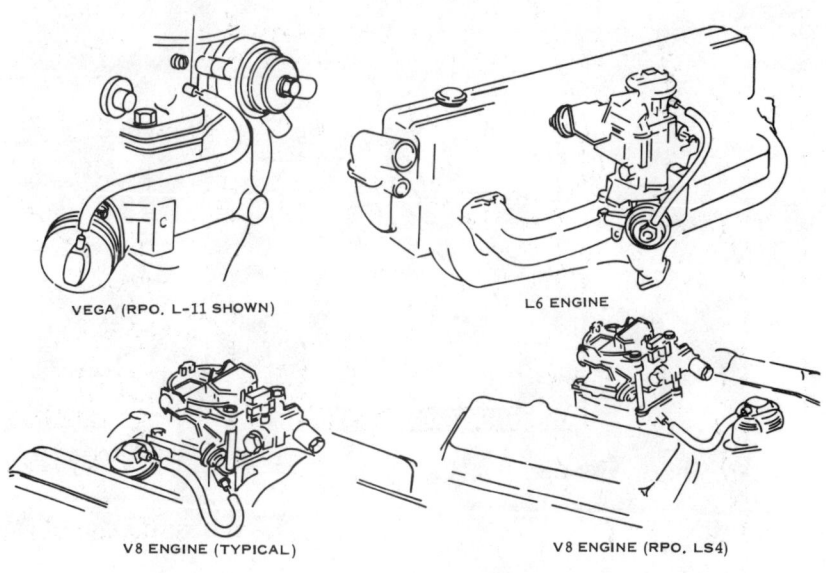

Position of the EGR valve on all Chevrolet engines.

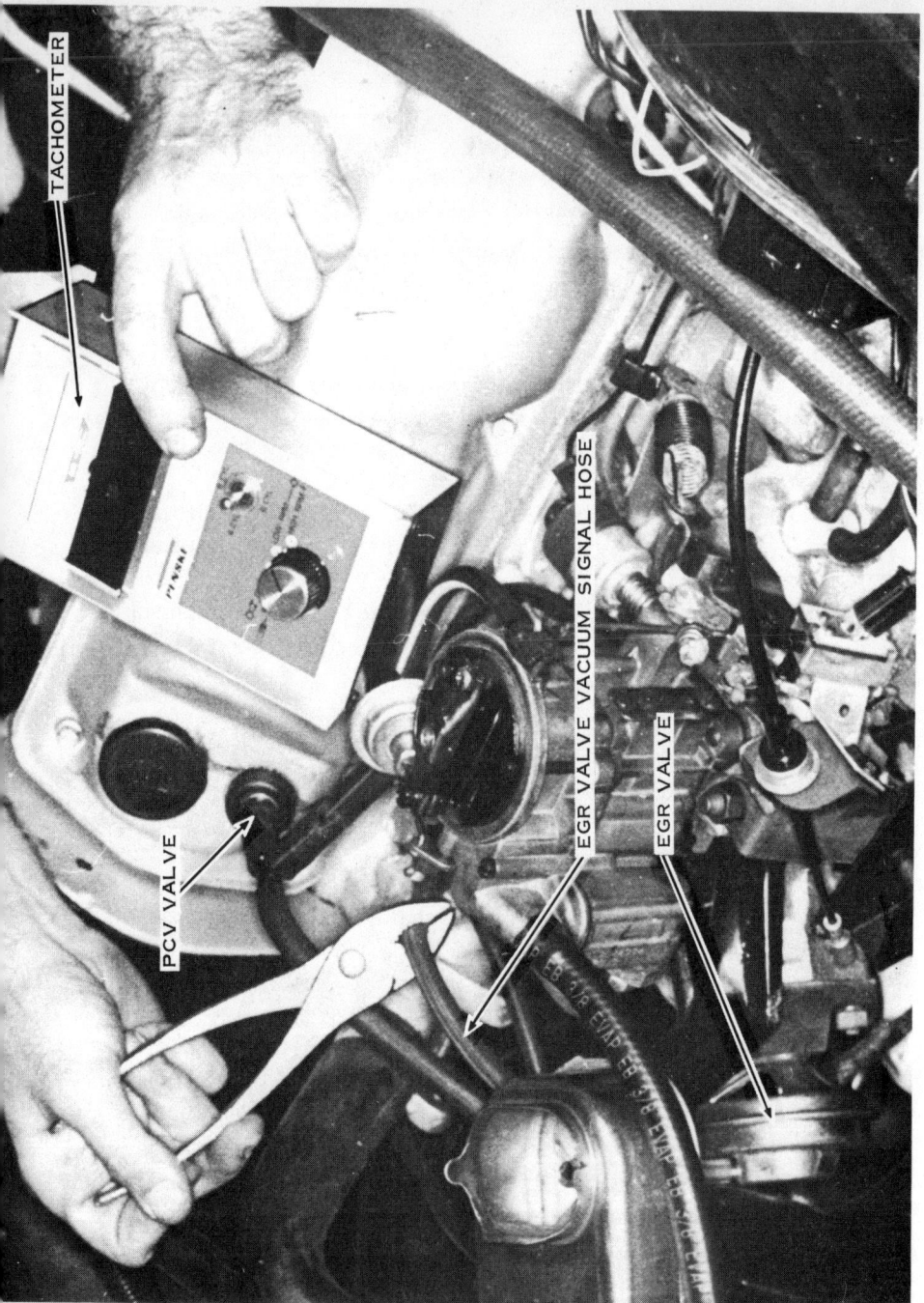

One way to test the EGR valve and its control circuit is to clamp shut the EGR valve vacuum signal hose with a pair of pliers, and then run engine speed up to about 2,-000 rpm. Note the tachometer reading, and then release the clamping hose pressure. Engine speed should drop at least 100 rpm to indicate the passage of exhaust gas into the intake manifold as the EGR system becomes effective.

CHAPTER 3
vehicle maintenance for economy

Vehicle weight is the greatest enemy of gas mileage during low-speed driving, and wind resistance takes over as the culprit at higher cruising speeds. It takes energy to push a car on level ground and even more energy to push it up a slight incline. And, the heavier the vehicle, the more work is required to get it moving.

A great power loss occurs from wind resistance at Freeway speeds. The effort required to shove all of that air out of the way increases as the cube of the vehicle's speed. That means that if you double your speed, you require eight more times as much horsepower to overcome air resistance ($2 \times 2 \times 2 = 8$). That's a lot of horses doing nothing but pushing air out of the way! And, too, anything that destroys the smooth flow of air over the vehicle's surface (aerodynamic drag) increases the wind resistance considerably.

Rolling resistance increases with speed, but not as dramatically as does the air resistance. Anything that adds to rolling resistance makes extra power demands on the engine, and this costs extra fuel.

All motorists spend money for gasoline and oil and, if you know what it's all about, you can save some bucks here.

Just keep chipping away, and your operating costs will be reduced. Possibly not dramatically by any one thing but, item by item, they'll add up to a total which may pleasantly surprise you.

VEHICLE WEIGHT

Obviously, the greatest weight factor is the size of the car. If you're in the market for a new car, then you can consider the following:

In a recent News Release, Ford Motor Company's president, Mr. Iacocca said, "Everybody is worried about the big gas guzzlers." "I am too," he asserted, "but what worries me more are the so-called little gas guzzlers like Pinto. Our Pinto came out at 2,-100 pounds, but in three years got fat. We put on 33 pounds for things like better steering. We also added 307 pounds to comply with government safety, emissions and damageability standards,

This Mustang II carries this load continuously, so the salesman cannot expect to get top mileage.

and to make the car run after we added the federal equipment. Gas mileage took a nosedive, from 25.0 to 21.0 miles per gallon—a loss of 19 per cent. This fall we'll add 75 more pounds for government standards..." He further made the point that, "What we need today are laws that strike an intelligent compromise between our need and desire for clean air and our absolute need to make the best possible use of our available fuel supplies."

EXTRA WEIGHT

If you decide to keep your present car, it's the extra weight that you can control, and you must "take it off," the extra weight, that is!

For each additional 100 pounds that you lug around, you are wasting 0.2 mpg. That's $9.25 per year, 77¢ per month, or 3¢ per day. Your air conditioner weighs about 100 pounds, even if its not being used. Of course, you can't just leave it home, but are you carrying around other unnecessary items of similar weight? How about checking your luggage compartment right now?

How about the weight of your gas tank? If you run it full at all times, you're lugging around approximately 20 gallons, which weigh 120 pounds and this is costing you about one buck per month just for trucking fees. Keeping the tank half full costs you only 50¢ a month for these unnecessary costs, which is a saving of $6.00 per year.

You can't expect to get good gas mileage when you're dragging around a boat.

GAS GUZZLER'S GUIDE

The moral of this story is to clean the luggage compartment of all unnecessary items and run the vehicle with only half a tank of gas, but don't let it run dry, or the emergency service call to your auto club will nullify your savings.

WIND RESISTANCE

If you stick your arm out of the window of a fast-moving vehicle, you'll feel the resistance of air against your hand. Turn your palm

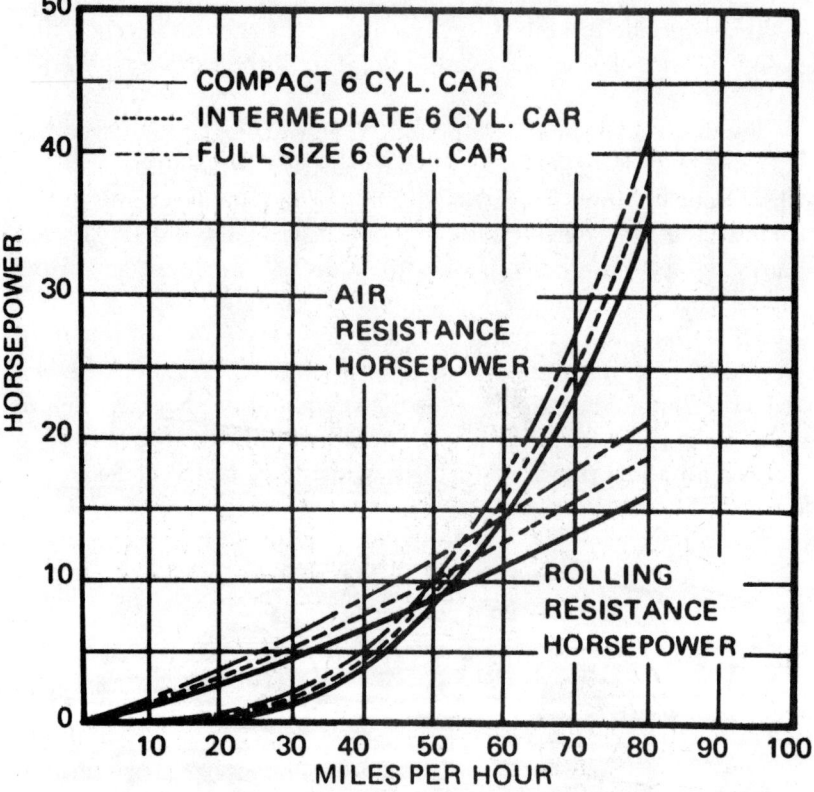

This graph plots the horsepower demands of a compact, intermediate, and full-size car in regard to air and rolling resistance. Note the steep curve for the horsepower required to overcome air resistance as compared with that needed to overcome rolling resistance.

to the front (maximum area), and the pressure increases. Turn it so the side of your hand faces the front, and the resistance decreases, but it never fades completely away. The analogy is that the more streamlined the vehicle, the less power required to "drive" it through the air.

For those mathematically minded, the algebraic formula is:

$Hp = DVAeV^3$.

Where:

K = A constant
D = Vehicle's drag (aerodynamic efficiency)
A = Frontal area
e = Density of the air
V = Speed

The drag efficiency relates to the aerodynamic "cleanliness" of the vehicle. If its shape is smooth, then the efficiency is relatively high. Such things as an open window, a luggage rack, or a vinyl roof change the smooth flow of air over the vehicle and this increases the

A permanently installed rack adds to the air resistance by destroying the aerodynamic shape of the vehicle.

drag. This is why a knowledgeable pickup truck driver will secure the rear compartment door (tailgate) in a horizontal position when he is driving without a load.

Note that the velocity (speed) figure is cubed (V^3) in the formula, and this is where these small Hp losses at low speed become so significant at higher speeds.

ROOF AND BICYCLE RACKS

A permanently installed rack adds to the air resistance, especially when it is loaded. You can save on fuel if you store your luggage inside of the vehicle to reduce wind resistance. However, if you do need to carry luggage on the roof at times, then buy a temporary type, which can be removed when not needed.

VINYL TOP

If you're in the market for a new car, note that a vinyl top offers enough added air resistance to penalize you at least 1 mpg at Freeways speeds, and this costs about $50.00 per year in added fuel. Is it worth it?

According to a recent Chrysler news release, "Vinyl roofs were preferred by 62% of Dart owners, making them six times more popular in 1974 than in 1967 when they were introduced. Move up to the full-sized models and the vinyl option is found on almost two-thirds of the cars. In the luxury field, Imperial has a vinyl roof as standard equipment." It is understandable why people who buy Imperials are unconcerned about gas mileage, but the Dart owner is defeating his own purpose.

ROLLING RESISTANCE

Rolling resistance increases with speed. It takes about 7 Hp to overcome the rolling resistance of a compact car at 40 mph and 15 Hp at 70 mph. That's a 200% increase. It's never possible to reduce rolling resistance completely, but you can minimize its effects by reading the next few sections on the factors which influence rolling resistance.

Remember, item by item, we must reduce all unnecessary drains on our pocketbooks to achieve the goals established by this book.

TIRES

The type of tires and their inflation pressure determine, to a remarkable extent, the rolling resistance of the vehicle, even at low speeds where air resistance is not a factor. If you're driving mostly at low speeds and your gas mileage is down, you can realize an improvement by making some of the following tests.

Possibly the most important single fact to minimize tire rolling resistance (and also to increase the life of the tire) is to maintain the proper air pressure. Air pressure lets your tires stand up straight to carry the designed load. When the pressure is too low, the side walls bulge and the tire's "footprint" increases. The added rolling resistance and reduced tire life can inflate your operating costs considerably.

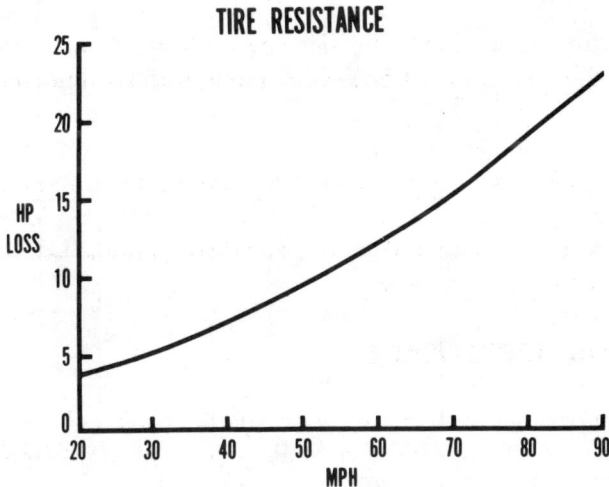

Graph showing the rolling resistance loss with speed. Note that it takes about 7 Hp to overcome the rolling resistance at 40 mph and twice as much at 70 mph.

This is the type of decal secured to your car's door post or glove compartment door. Study it and determine just how much pressure is required for your own vehicle and load.

RECOMMENDED TIRE INFLATION PRESSURES (Pounds Per Square Inch Cool)				
Models	Standard Inflation For All Loads Including Full Rated		Optional Inflation For Reduced Loads	
All Except Station Wagons Equipped With:	1 to 6 Passengers + 200 lbs. Luggage (1100 lbs Load)		1 to 5 Passengers (750 lbs. Maximum Load)	
L-6 or 350 V-8 Engine and 455 V-8 Engine	Front	Rear	Front	Rear
	26 lbs.	28 lbs.	24 lbs.	24 lbs.
Station Wagons	1 to 6 Passengers + 300 lbs Luggage (1200 lbs. Load)		1 to 5 Passengers (750 lbs. Maximum Load)	
	Front	Rear	Front	Rear
	24 lbs.	32 lbs.	22 lbs.	26 lbs.

Use of optional inflations is allowable only with a reduced load (one to five passengers). When operating at loads greater than the optional reduced load, the inflation pressure must be increased to the standard inflation for full rated loads.

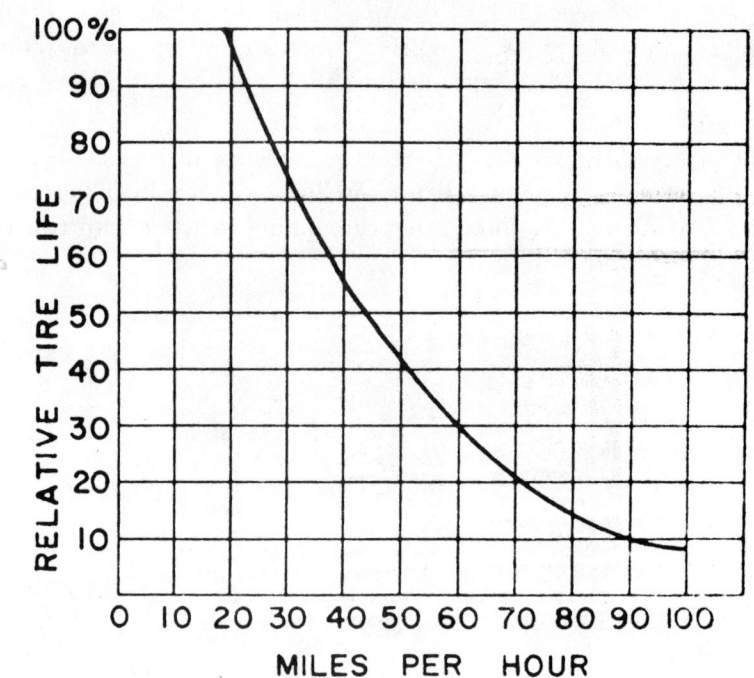

Tire life, too, decreases rapidly with increased speed.

TIRES 111

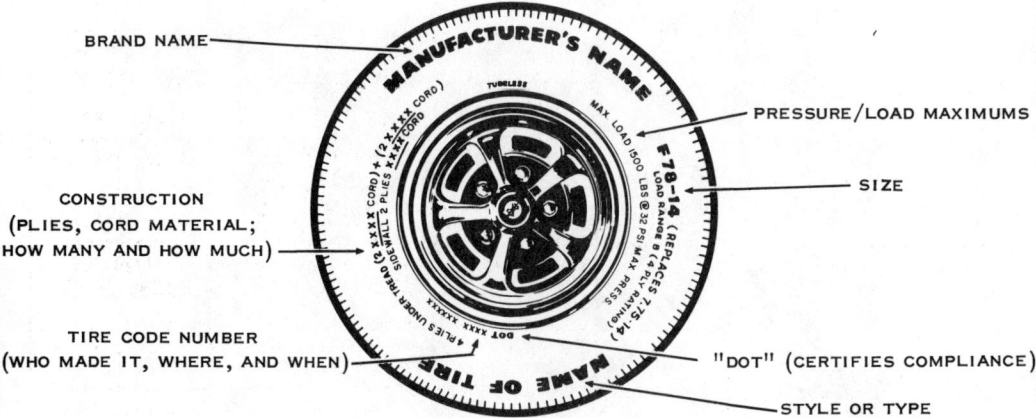

A tire has many code numbers, and this drawing tells you how to read them. Note that the maximum air pressure on this tire is listed as 32 psi, and this must not be exceeded at any time; otherwise, you will weaken the sidewalls.

All new cars are required by law to have a decal secured to the door post or glove compartment door (as well as being printed in the owner's manual) with the recommended tire pressures. If you were the owner of the car with the accompanying decal, you should normally carry 24 psi (pounds per square inch) pressure in all four tires if you regularly use the car for commuting to work with two or three passengers.

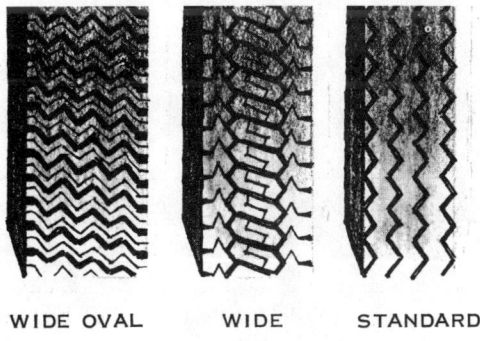

Wide-oval tires have a much larger footprint than normal treads and, therefore, have greater rolling resistance.

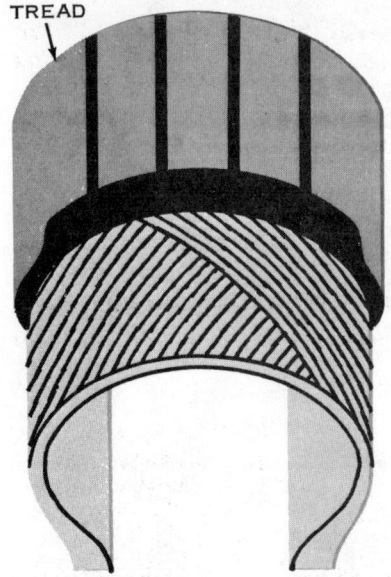
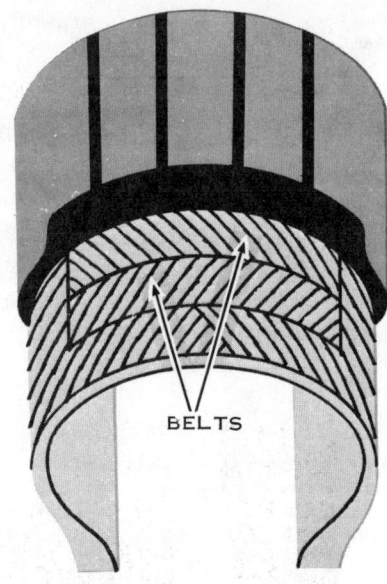
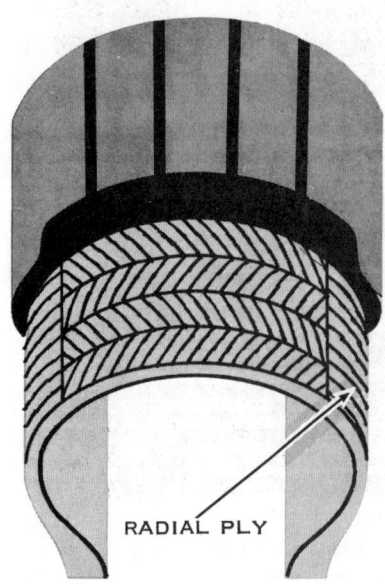

The bias ply tire (upper left) is the conventional one that has been used since 1920. The cords criss-cross at an angle, called the "bias angle," which is usually 30-40°. This tire tread squirms a good deal building up heat, but the ride is much softer than the other types.

The belted bias tire (upper right) has two extra layers of fabric or "belts" under the tread. The cords in the belt run at an angle of about 25° to the centerline. This type of construction adds stiffness to the tread, which minimizes squirm and heat build-up. The ride is harsher, but the gas mileage is better than with the bias ply tire.

The radial tire (left) has cords in the body running at right angles to the centerline of the tire. Several belts are used between the plies and tread, and this provides the least rolling resistance of any tire.

However, if you put your whole family of six in the car for a vacation trip, along with the family luggage, you should raise the front tire pressures to 26 psi and the rears to 28 psi.

Check your tire pressure monthly with a good gauge. Service station gauges are very often inaccurate; they can be off by 2-15 psi, and this can affect your rolling resistance as well as tire life. Always check your tire pressure cold, never after driving for any distance. After driving on a long trip during a hot day, you might find that your tire pressure has built up as much as 5 psi. Don't bleed off the extra pressure, because then you will be running underinflated when the tire is cold.

Picture of a tire's footprint taken through glass.

REDUCING TIRE ROLLING RESISTANCE

The higher the tires are inflated, the lower the rolling resistance and the better the gas mileage. However, the greater the inflation, the harsher the ride. So you must make a trade-off here and determine just how much ride quality you are willing to sacrifice to save some money. Try the basic test outlined in the first chapter with the tires overinflated by 2 psi and then by 5 psi. As you road-test the vehicle, determine accurately the improvement in mileage and the loss in riding quality. Adjust the pressure as high as you can stand

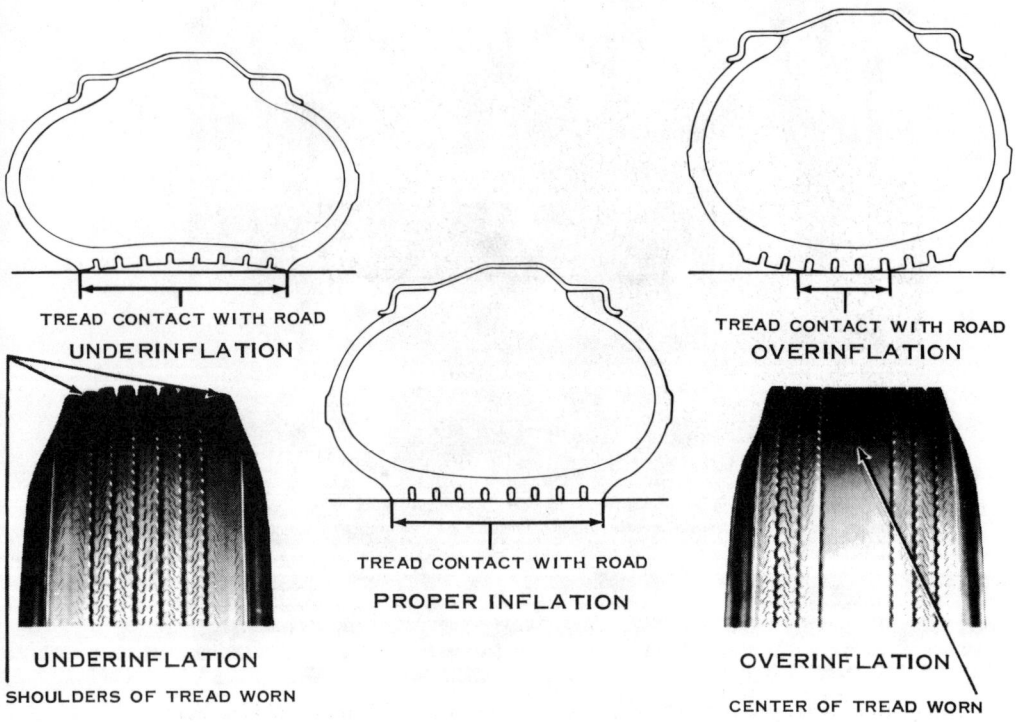

Tire operational conditions and resulting wear.

it; the thickness of skin on your bottom will be the determining factor here.

If you are in the market for new tires, avoid the wide-oval type because of its larger "footprint," which creates greater rolling resistance. Also, they ride much harsher than a conventional tire.

Note, too, that radial-ply tires offer a minimum of rolling resistance, and their use can increase your gas mileage by 5%. You can use the $2.50 monthly savings to pay for the tires. However, radial-ply tires do ride harsher at low speeds than do bias-ply tires. Try them out first, if you can work such a deal. You'll have to trade riding quality for bucks, whichever suits your fancy! New cars, with radial-ply tires as standard equipment, have the suspension tuned for their riding characteristics, but this is not true when you install such tires on older models. The results can be a surprise—sometimes unpleasant.

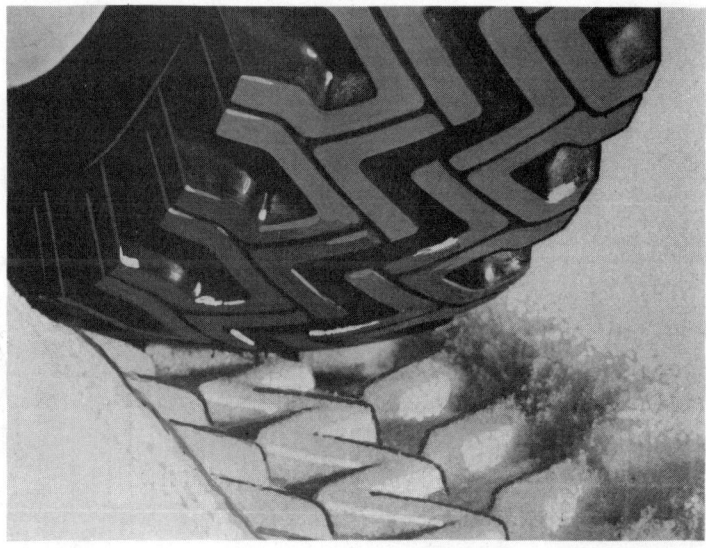

Snow tires have special treads to get you rolling under difficult conditions, but their rolling resistance is much greater than any other type. There goes your gas mileage!

116 GAS GUZZLER'S GUIDE

Excessive toe-in causes this type of tread wear.

The condition of the roadway affects your gas mileage to a considerable extent.

WHEEL ALIGNMENT

Anything that adds to rolling resistance increases fuel consumption. Improper front wheel alignment can increase fuel consumption by 0.3 mpg, and this means that you're wasting about $14.00 per year in gasoline, as well as scuffing off your tire treads, which can make you purchase a new tire long before it should be necessary. If you're scraping rubber, it will show up as rough, feather-like edges on one side of each ridge. Run your hand lightly across the tire face. If you feel sharp edges going one way but not the other, then it is time to get your front wheels aligned to save money, both on gasoline and tires.

ROAD SURFACES

The condition of the roadway will influence your mileage by affecting the rolling resistance of your tires. If you do have a choice, you may be able to select an alternate route that does not penalize you when you consider the following factors: A patched asphalt roadway can reduce your mileage by about 15%, a loose gravel or dirt road by 35%, and a sandy surface by as much as 45%.

DISC BRAKES

Disc brakes are superior to drum brakes in many ways but they do increase rolling resistance because the pads are always lightly touching the rotors. The effect can be multiplied many times over if the driver has the bad habit of resting his foot on the brake pedal. Don't laugh, many drivers do just that.

Drum-type brakes can add to rolling resistance if they are not properly adjusted. To check this out, run the vehicle on a level road at about 30 mph. Shift into neutral and allow the car to coast to a stop, which must be smooth to indicate no extra rolling resistance. If there is a small, but perceptible jerk at the end of the stop, then you have extra drag in the braking system or in the axle bearings, which should be investigated by a competent mechanic.

BUYING GASOLINE WISELY

Aside from the savings you can realize in reducing your consumption of fuel by using this book, you should learn how to save money by buying the right kind of gasoline. Such terms as premium, high test, regular, low lead, and unleaded gasolines add to the confusion. In addition, the octane numbers posted on the pumps, be they "Motor", "Research", or "Road", or an average of them, do nothing to help the average driver. Really, all that you have to remember is to buy the lowest octane number (price) gasoline that will operate your engine properly.

With gasoline costs varying about 15¢ between premium (high test) and economy regular brands, you can realize a yearly savings of up to $90.00 by selecting wisely. In any event, buy the gasoline

Type of sticker attached to many new vehicles to inform the owner just which gasoline should be used.

with the lowest posted octane number that will run your engine without detonation.

If you have a 1975 vehicle with a catalytic converter, you have no choice and must buy unleaded gasoline to avoid "poisoning" the converter, and so the rest of this discussion is not for you. If you have an older car with a high-compression engine (through 1970), you must use premium (leaded) gasoline to avoid detonation. If your car has been manufactured between 1971-74, you do have a choice of octane numbers and price. Naturally, the lower the price per gallon of gasoline you buy, the less your operating costs. In a word, "Don't overbuy!"

	Economy		Regular			Mid Premium			Premium		
Federal octane on pump	87	88	89	90	91	92	93	94	95	96	97
Research octane	91	92	93	94	95	96	97	98	99	100	101
Amoco	Blue		Regular*	Regular*					Super Premium		
Arco	Clear			Arco					Supreme		
BP			Regular						Super		
Chevron		Unleaded	Chevron						Supreme		
Citgo			Regular						Premium		
Crown			Regular*	Regular*					Extra		
Exxon				Regular	Plus				Extra		
Gulf		Gulfcrest	Good Gulf						No Nox		
Hess				Regular					Premium		
Mobil			Regular	Special					Premium		
Phillips 66			66						Flight Fuel		
Shell				Shell Regular	Super Regular				Super Shell		
Sunoco			190	200			220		240		260
Texaco				Fire Chief						Sky Chief	

*Varies; go by rating on pumps in your area.

Guide to octane ratings of popular gasolines. This chart was developed by the Office of Consumer Affairs.

120　GAS GUZZLER'S GUIDE

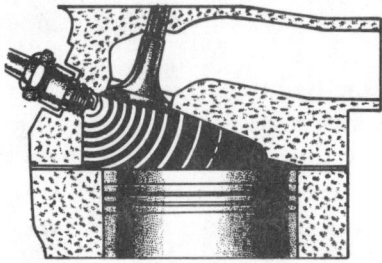 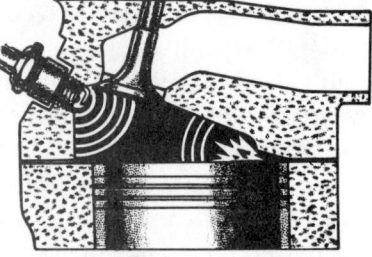

NON-KNOCKING COMBUSTION　　KNOCKING COMBUSTION

Detonation (pinging) is caused by uncontrolled combustion. It can be minimized by using a gasoline with a higher octane rating.

You can use regular gasoline without pinging when driving in mountainous areas where the air is thinner and your engine works at a lower volumetric efficiency.

OCTANE NUMBERS

The octane number of a gasoline is its ability to resist detonation (pinging), which are sharp metallic noises resulting from uncontrolled combustion. The higher the octane number, the better the gasoline resists knocking. Buy the lowest octane number (cheapest) gasoline you can, provided your engine does not ping; that's the secret.

That seems simple enough and you can find the octane number requirement for your new car in your owner's manual, but it becomes more complicated when other factors enter the picture: carbon accumulation in the combustion chambers of older cars (which can require a 4-6 higher octane numbered gasoline), variations in engine and air temperatures or humidity, and changes in altitude (where the thinner air lowers the efficiency of your engine).

Your problem can be simplified somewhat if you buy a single name brand, such as Chevron or Texaco, because you merely have to use the cheapest type of their gasoline that does not knock. If necessary, mix premium and regular to get the type you need. Possibly you can use a mixture of 50/50 or 25/75, whichever suits your needs for proper engine operation and lower initial costs.

This "gassy" subject cannot be closed without a discussion on the addition of lead (tetraethyl), which is added to gasoline to increase the octane number and reduce its sensitivity to engine knock. Lead also performs the important function of lubricating the valve guides and minimizing valve seat burning. Therefore, if you have a car without a catalytic converter, you should always buy some of your gasoline with lead in it. If you are buying no-lead gasoline, then every fourth fill-up should be one with leaded gasoline to minimize valve wear.

If you have a vehicle powered by a high-compression engine which requires premium (leaded) gasoline at all times to minimize knocking, then you can switch to regular gasoline whenever you are operating at higher elevations, say about 2,000' above sea level, where the thinner air lowers the working pressures (volumetric efficiency) and sensitivity of your engine to detonation.

BUYING OIL WISELY

Like any other comodity in the highly competitive automotive aftermarket, lubricating oils can be purchased much cheaper at auto supply stores than at a filling station. If you are inclined to save money this way, add your own oil.

If a service station attendant wants to check under the hood, step out to watch him. Make sure that he pushes the dipstick all the way in after wiping it off. If he doesn't get it all the way in, then the oil level on the dipstick will show lower than it actually is. If you do get "short-sticked" and add unneeded oil, then you will be operating with too high an oil level which, in extreme cases, will allow the lower end of the connecting rods to dip into the oil to churn and aerate it so that it foams out of even normal seals. On the other hand, you can carry the oil level up to one quart low without ill effects and this minimizes excessive oil churning.

Engine oils with a lower viscosity decrease friction and, therefore, are highly desirable as a means of increasing gas mileage. The lower the code number, the lower the viscosity. Multi-grade SAE 10W-30 oil has a lower viscosity when cold than does the single-grade SAE-30 oil. Both of them have the same protection (30) when hot. Don't use an oil of higher viscosity than is recommended in your owner's manual, since heavier oils increase friction and decrease gas mileage.

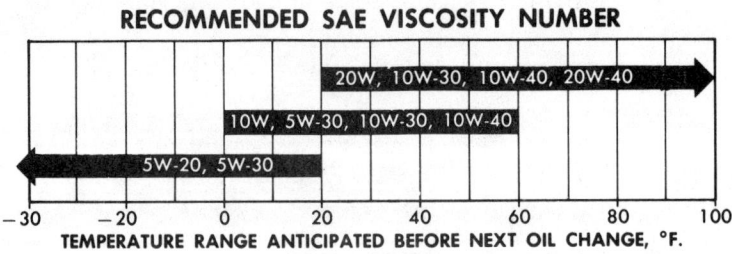

Oil viscosity recommendations according to ambient air temperatures.

CHAPTER 4
how to drive for economy

Keeping your engine tuned and proper maintenance of tires, brakes, etc., will help, but you must watch your driving habits too. Look how the fuel consumption increases and miles per gallon decreases with speed. We've already talked about this in Chapter 3. Driving 55 mph may take 20 or 30 minutes longer than hitting 65. But at 65 mph, you're losing 2 or 3 miles per gallon economy. (And maybe you'll get a ticket!) And what are you going to do with that 20 or 30 minutes?

Your driving skill determines, to a great extent, your ability to get good fuel economy. A "heavy foot" is a drain on your pockedbook.

Then there are those who don't drive at a steady highway speed. Speeding up, slowing down, speeding up, slowing down ("sea-sawing"), and this can increase fuel consumption considerably. And how about those who drive with the left foot resting on the brake pedal? Look at the extra drag they're putting on the engine. Watch those prolonged idling periods, too. They're murder on mileage.

All of these gas mileage "depressants" are fully discussed in the first section of this chapter under the heading "Improving Your Driving Skills." The second section includes information about some add-on equipment that can favorably affect your fuel bills.

Both of these topics continue our efforts of "taking it off," and we know from experience that they'll work. However, if you haven't

Today's stop-and-go driving is very wasteful of fuel.

succeeded in getting your operating costs down after reading this chapter, then you'd better start rereading the book again, because you just didn't take me seriously enough in getting down to the "bare" facts.

IMPROVING YOUR DRIVING SKILLS

The most important single element in determining the fuel consumption of a car is the driving techniques of the individual behind the steering wheel (that's you!). One authority declares that a knowledgeable driver can obtain at least 30% better mileage than the average driver and 50% better than a poor one.

If you already are an average driver and do utilize the driving tips described in this book, you can be saving about $15.00 per month. If you're a poor driver (in regards to our economy driving tips), then you're going to save $25.00 per month. "How can I save these bucks," you ask? Read on!

Jack-rabbit starts can increase fuel consumption at least 2 mpg in city driving, as well as stripping off your tire treads.

126   GAS GUZZLER'S GUIDE

STARTING A COLD ENGINE

An internal-combustion engine is least efficient when it is started on a cold morning. Because the gasoline is not in vapor form, an automatic choke has to restrict a great deal or air flowing through the carburetor so that only enough is allowed in to mix with the partially vaporized fuel for a combustible mixture needed to start the engine. This means that a great deal of liquid fuel is being pumped through the cold engine and never burned (wasted).

In addition, consider that the internal-combustion engine develops its power by the process of combustion, and it requires a lot of fuel to raise all of that cast iron to operating temperature. Everytime you let the engine cool down, you must consume this extra (wasted) fuel to raise the temperature back to its normal operating range.

It's going to take a lot of energy (gasoline) to start this engine and warm all of that cast iron to operating temperature. But even in sunny Florida, it takes a lot of energy to heat the engine to operating temperature every time you let it cool down.

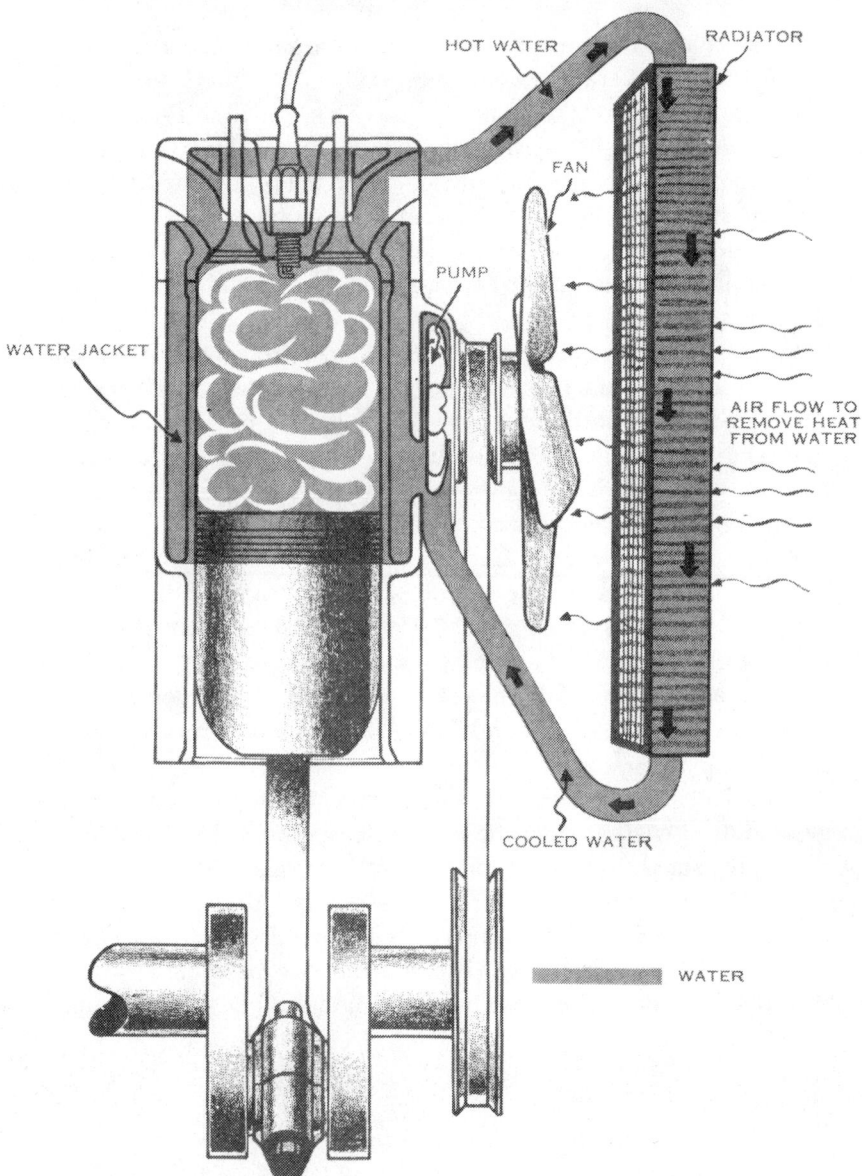

The internal-combustion engine is cooled by air. The coolant passes around the water jacket, picking up heat, and is pumped to the radiator and cooled by air flowing through it.

To start a cold engine properly (with a minimum of wasted fuel), depress the accelerator pedal once (about half way) to set the automatic choke, and then turn the ignition key to the START position. Don't pump the accelerator pedal while cranking the engine; the automatic choke is designed to provide the proper air-fuel mixture for starting under all temperature conditions. If you do pump the accelerator pedal, you're adding extra (wasting) fuel and can slow ignition because of the raw fuel you pumped through the carburetor.

When the engine does start, don't waste gasoline by letting it idle to warm up. Just start driving, but do it at a conservative speed of about 20-25 mph until the "cold" warning lamp on your dash goes out. The engine will warm up quicker while it is under load and, too, you will be covering some of the trip mileage so that the warm-up fuel is not entirely wasted.

A cold engine averages about 6 mpg, depending on the ambient air temperature, size of the engine, and the weight of the vehicle. The same powerplant can average about 12 mpg after about 15 miles of driving. If your driving consists mostly of short trips, then you're wasting about 50% of your gasoline or, said another way, you're buying an extra 40 gallons of fuel each month at a cost of $24.00.

Then, too, frequent engine starts, especially when cold, draw a great deal of current from the battery to operate the starter. After the engine starts, the alternator must recharge the battery, and this can lower your gas mileage from 0.5-0.9 mpg, depending on just how much current must be returned to the battery. Naturally, your alternator is not working at maximum capacity all of the time, and this loss depends entirely on just how many cold engine starts you make.

Thermostat

Your engine coolant should warm to operating temperature (red warning light on the dash goes out) within 3/4 mile after a cold start. If it takes longer than that, the coolant thermostat may be defective, and your engine can be operating too cool. This can result in an operating loss of 1-2 mpg, for a cost of 50-90 dollars per year.

COLD ENGINE STARTING

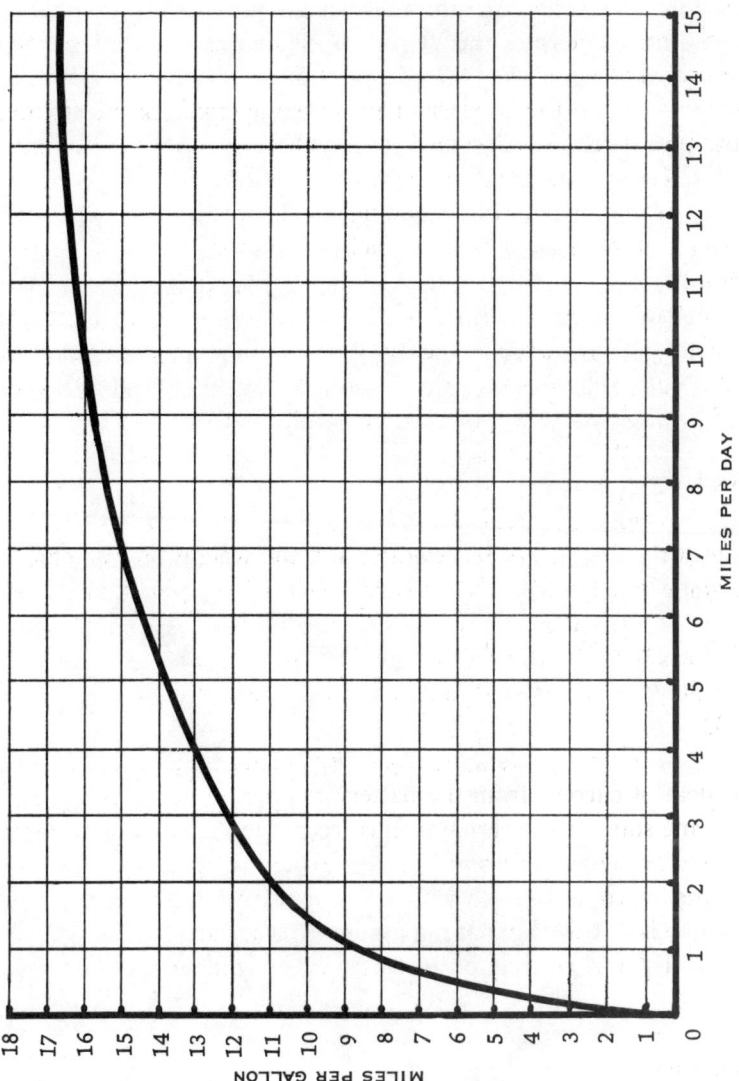

Your fuel economy during the uneconomical warm-up period is about half of what you can get after the engine is warmed to operating temperature. This graph is based on tests made during an ambient air temperature of 10°F.

If you do have to replace the thermostat, use one of an equal or higher rating to keep the engine operating at a more-efficient temperature.

PREPLANNING

Possibly, you can't do too much about this wasted fuel, except to keep the vehicle stored in a closed garage overnight to minimize this heat loss. Preplan your trips by combining short shopping and commuting trips to reduce the number of cold starts and the miles traveled each time. Patronize the shops in your neighborhood as much as possible to reduce your wasted fuel costs and to lower pollution, too.

EXCESSIVE IDLING

An idling engine can consume almost a cupful of gasoline every

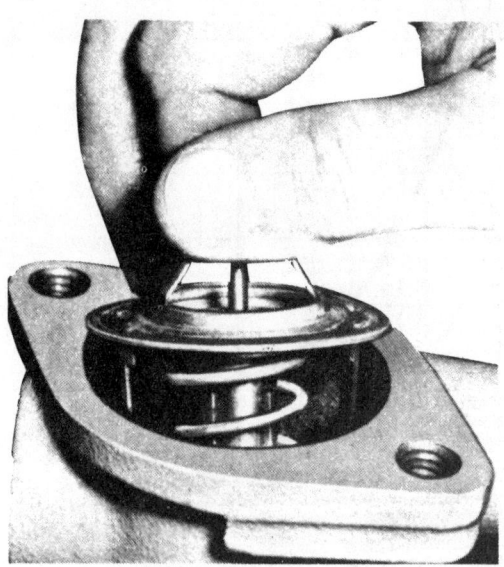

The thermostat is used to block the passage of coolant to the radiator until the block coolant is heated to operating temperature. At this point, the thermostat opens and allows coolant flow to the radiator for cooling purposes.

IMPROVING YOUR DRIVING SKILLS 131

six minutes. Therefore, if you're waiting for someone, turn off the engine if the wait is to be for any length of time.

GETTING UNDERWAY

Start slowly and accelerate gently whenever possible. Hot-rod starts and jerky acceleration can increase fuel consumption by two mpg in city traffic, and this can cost you $100.00 a year. Worthwhile? You tell me!

Starting out slowly and accelerating easy through the gears can be a real money saver. If you have a vacuum gauge mounted on the dash, keep the needle in the economy zone and never let it drop below 10"Hg of vacuum, which will keep your carburetor power jet from opening. That can be a real saving and the amount depends entirely on how many times you have to stop and start up at traffic lights and stop signs.

He's not trying to get good gas mileage. And, if you drive that way, you're not going to get any gas mileage, either!

PACING TRAFFIC LIGHTS

Try to pace traffic lights. When you see a red light ahead, slow up and try to adjust your speed so that the light changes to green as you get there. Good drivers are constantly seeking out the condition of traffic lights as far ahead as possible and adjusting their speed accordingly. And this means avoiding stops, each with its gas-consuming start-up. It's difficult to estimate the savings here, but it can be considerable, depending on your skill in avoiding unnecessary stops. Make a game of it, and practice constantly to sharpen your skills. Give yourself a quarter for each red light you avoid and penalize yourself 50¢ for each one you get caught on.

Pacing traffic lights can be a rewarding experience. Make a game of it to sharpen your driving skill.

BE AN "OLD SMOOTHIE"

Everytime you accelerate, you have to change the speed of your vehicle, and this means shoving along 2 tons of metal. Try pushing it once and you'll appreciate the effort required to get it moving. Just as you must expend muscle power in pushing the car, it is necessary to burn fuel for the engine to develop the power for moving and accelerating the vehicle.

To minimize this loss, it is essential that you drive at a steady pace as much as possible, neither accelerating or decelerating, whenever traffic conditions permit. To this end, drive as if you have an egg on the accelerator pedal so that you are always conscious that

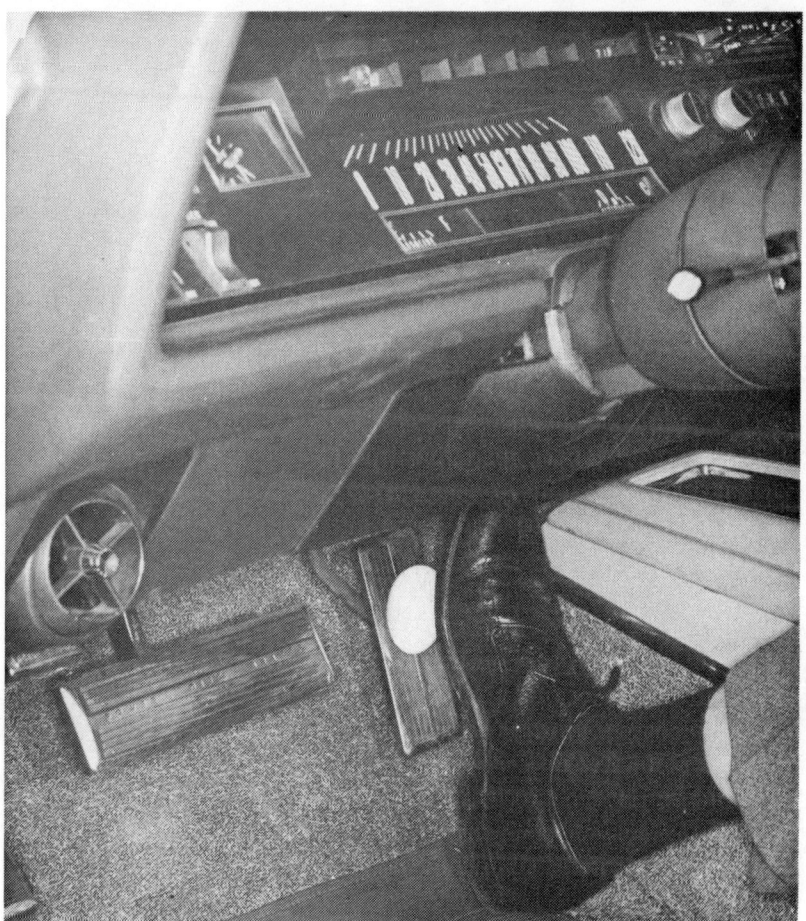

If you drive as if you had a raw egg between the sole of your foot and the accelerator pedal, you'll get much better mileage.

added pressure could break the shell and splatter the contents all over the place. If you make an effort to think about this often, you'll find that it soon becomes part of your driving habits and you'll become an "Old Smoothie." It's amazing just how much fuel you can save this way, but this will be in inverse proportion to the thickness of the callus on the sole of your right foot. If you're thin-skinned, you'll save a "bundle."

It might be of interest to know that several years ago Mobile Oil Company conducted an annual economy run. A technique used by several expert drivers, who placed at the top of the list, was to remove the right shoe so that the foot was more sensitive to the feel of the accelerator pedal. That's really being an "Old Smoothie."

How to be an "old smoothie." Expert driver Micky Thompson removes his right shoe during a recent Fuel Economy Run in order to have a more-sensitive feel of the accelerator pedal.

DRIVING THE FREEWAYS

Driving at moderate speeds can increase your gas mileage considerably if a lot of your driving is on the Freeway. As your car's speed increases, so does wind resistance, and this becomes a really big factor at higher speeds. Note from the accompanying graph that it takes 3 Hp to overcome the wind resistance of a compact car moving along at 40 mph and 35 Hp to force it through the air at 80 mph. That's a 1200% increase in power demand just to push the air out of the way.

Note, too, that the rolling resistance increases with speed, but not as dramatically. At 40 mph, it takes 7 Hp to overcome rolling resistance and 15 Hp when you double the speed to 80 mph. That's a 100% increase.

If you're an average driver (economy minded, that is) and you normally travel the Freeway at 70 mph and reduce your speed to 55 mph, you can realize a fuel saving of about 20% and that's a monthly savings of $10.00. By reducing your speed to 50 mph, you can save another 5%, or a total of up to $150.00 per year, that's if you're always on the Freeway. Got the message—easy does it!

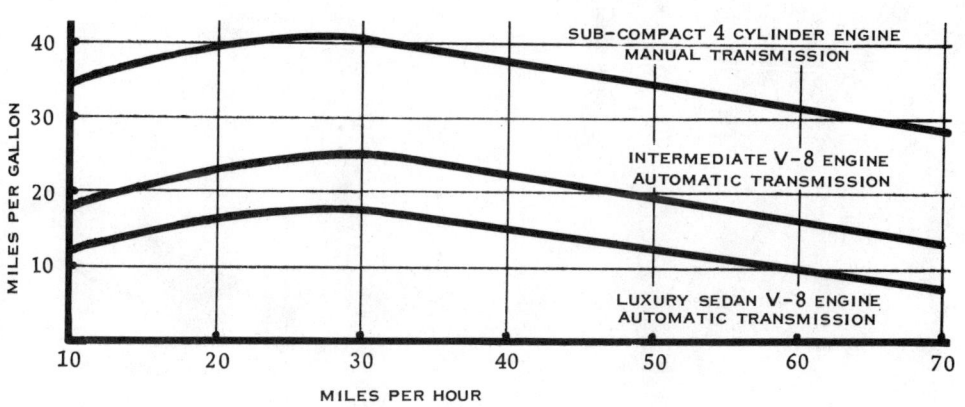

This graph shows the actual mileage obtained with three different sizes of vehicles. The tests were run on a level road at a steady rate of speed.

136 GAS GUZZLER'S GUIDE

DRIVING IN HILLY COUNTRY

Acceleration on level ground costs energy, but the bill goes up astronomically when you climb a steep hill. Just suppose the hill is 1,000' high. Do you realize that your engine is going to have to do work equivalent to lifting your 4,000 pound vehicle straight up 1,000 feet? That's right, straight up! It's the same as climbing a flight of stairs—you have to exert the same muscle-power to climb the stairs as to lift the same weight straight up. Climbing stairs appears to be easier, because the total is divided by the number of steps, but you must expend the same total muscle power.

And this can be applied to driving in hilly country. It's just going

You can take advantage of "old man gravity" by accelerating slightly on a downgrade if you see a steep hill ahead.

IMPROVING YOUR DRIVING SKILLS 137

The kinetic energy of speed is a tremendous force which is not always appreciated. This three-car pile-up gives an insight just how much energy a moving object stores. It takes a similar amount of energy (brakes or crash) to stop a fast-moving vehicle as it does to accelerate it in the first place.

This driver is primarily concerned about speed, not economy. It takes a lot of energy (gasoline) to get going. Look at the size of the hose they're using to fill the gas tank.

to take that much extra power (and fuel) to climb hills. Can we minimize this loss? Yes, by avoiding hills whenever you have a choice of routes. But a good technique to use in hilly country where you have no choice is to take advantage of the law of conservation of energy. If you are descending a hill and see a "biggie" ahead, then accelerate gently downhill to take advantage of "old man gravity," who adds kinetic energy (energy of position) to your car's speed dur-

When you do accelerate in hill country, make sure that you are not overdriving your ability to stop. The kinetic energy of speed is a force that you can put to good advantage if you keep it under control.

ing the downhill run. (This is the only time that it doesn't cost you extra fuel to increase your car's speed.) Now maintain the same or slightly more accelerator pressure during the uphill run, just enough to go over the top with a minimum of wasted energy. See, you've taken advantage of one of nature's laws to save precious gasoline.

ACCESSORIES THAT AFFECT GAS MILEAGE

This section will discuss those optional pieces of equipment that can increase your gas mileage: vacuum gauge, wooden block, and cruise control; and one that reduces your mileage, an air conditioner.

VACUUM GAUGE

A vacuum gauge measures the amount of vacuum in the intake manifold and this, in turn, is dependent on the load you are putting on the engine. Therefore, such a gauge can indicate gas-demanding driving conditions; when you are crowding the throttle too hard.

Normally, an idling engine's intake manifold vacuum is 15-18"Hg (inches of mercury). As you accelerate, you open the car-

The Oldsmobile optional fuel-economy meter is mounted as part of the dash instrumentation.

buretor throttle valve to allow more air-fuel mixture to enter the engine, and this lowers the vacuum. As you load the engine hard, the vacuum can drop below 8"Hg, and this action is used to open a power jet in the carburetor to enrich the air-fuel ratio in order to cool the combustion chambers during the time you are demanding that the engine work so hard for you. Any time that you open an extra jet in the carburetor, you're using more gasoline. Therefore, it is well-worth the 10-15 dollars that such a gauge costs to remind you whenever you're crowding the throttle too hard and opening that fuel-consuming power jet.

Some vacuum gauges have a dial numbered from 0 to 30, the

An optional fuel-economy meter (vacuum gauge) is available on many 1975 car lines.

VACUUM GAUGE 141

latter representing 30″Hg. Some of the gauges are color-coded, with a green segment as an economy zone (between 10 and 18″Hg). The lower part of the dial is divided into an orange segment marked FAIR and a red zone marked POOR. When the needle is operating in the red segment, you've opened your power jet and also your wallet. Like we said before, keep thinking that you have an egg on the accelerator pedal and too much pressure will crush it.

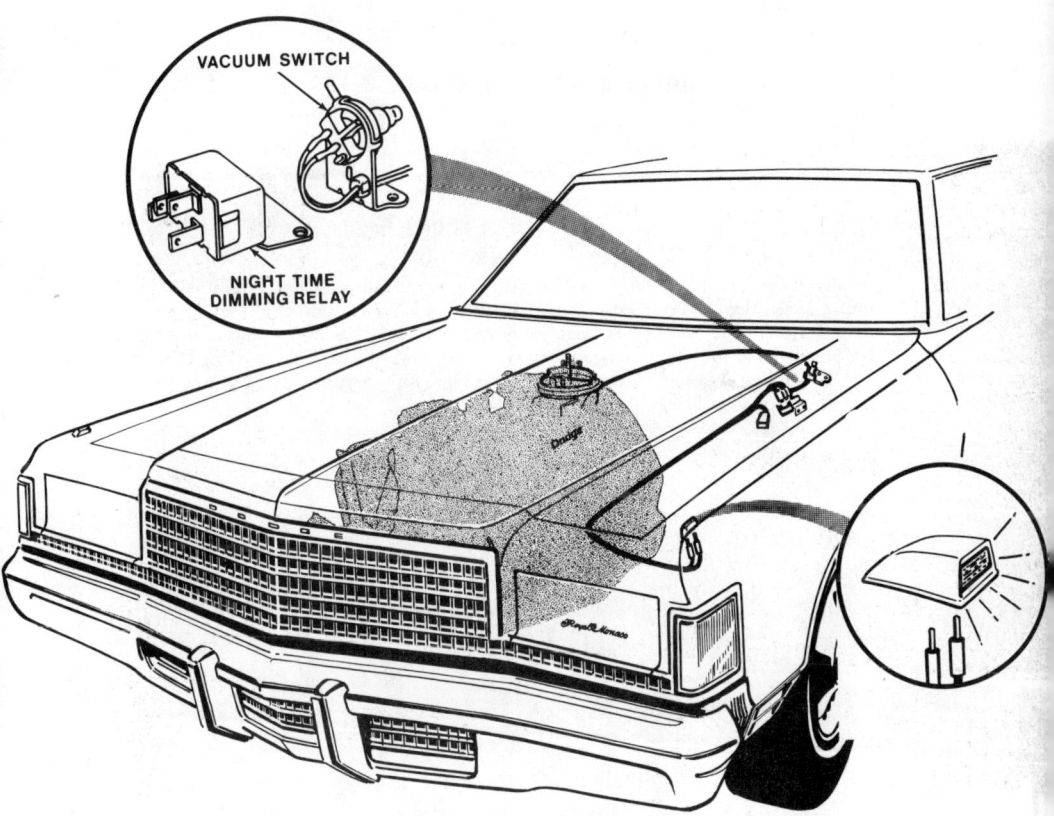

The Chrysler Fuel-Pacer system has a vacuum meter and a blinker light on the left-front fender for a visual warning to the driver that he is crowding the accelerator too hard.

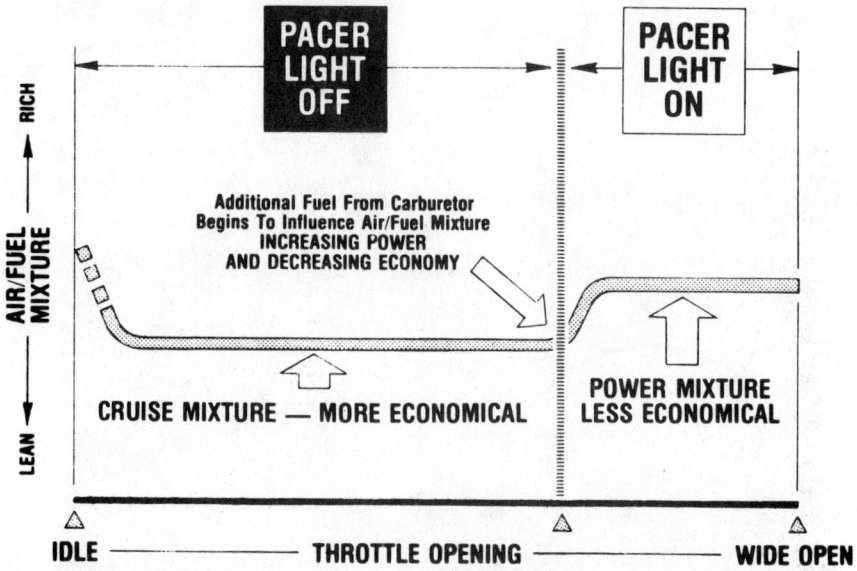

In the Chrysler Fuel-Pacer system, intake manifold vacuum operates a switch mounted in the engine compartment at about 4.5"Hg to light the lamp on the left-front fender. The amount of vacuum required to light the lamp correlates with the point at which the carburetor enriches the air-fuel mixture by opening the power jet. The light is designed to flicker at 5.5"Hg as a warning that the carburetor is about to deliver a richer mixture.

WOODEN BLOCK

An excellent way to minimize fuel-consuming speed variations is to position a block of wood under the accelerator pedal to restrict its full travel. My own gas guzzler, with a 440 cubic-inch powerplant, has an accelerator pedal that is positioned 3" above the toeboard, with a 2" travel. In my case, a 2-3/8" block of wood under the accelerator pedal limits its travel to 5/8", and this allows the vehicle to run precisely at 55 mph on level ground.

True, I can't downshift for passing, and my acceleration will never put me in the winner's circle at the drag strip, but I'm saving better than 50% of my gasoline bills in city traffic, and this can be $300.00 per year. Another thing, you'll never get a ticket for speeding on the Freeway, because you just can't exceed the speed limit, unless you do it going downhill.

CRUISE CONTROL 143

This accessory can result in at least a 4 mpg saving at Freeway speeds, which can be as much as $15.00 per month, if this is your main type of driving. You'll have to test various thicknesses of wood to determine just how much acceleration you are willing to sacrifice compared with the savings in "bucks." Remember that the wooden block costs nothing. Is it worthwhile? From personal experience, I know it is.

We're not fooling! An excellent way to save gas is to put a wooden block under the accelerator pedal. It's a cheap accessory.

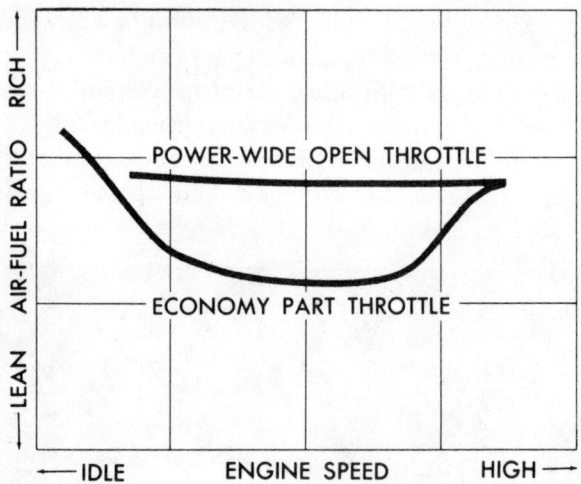

This graph of the carburetor air-fuel ratio shows why the Cruise Control is effective in reducing your gas consumption. It keeps your engine operating on the part-throttle (economy) curve.

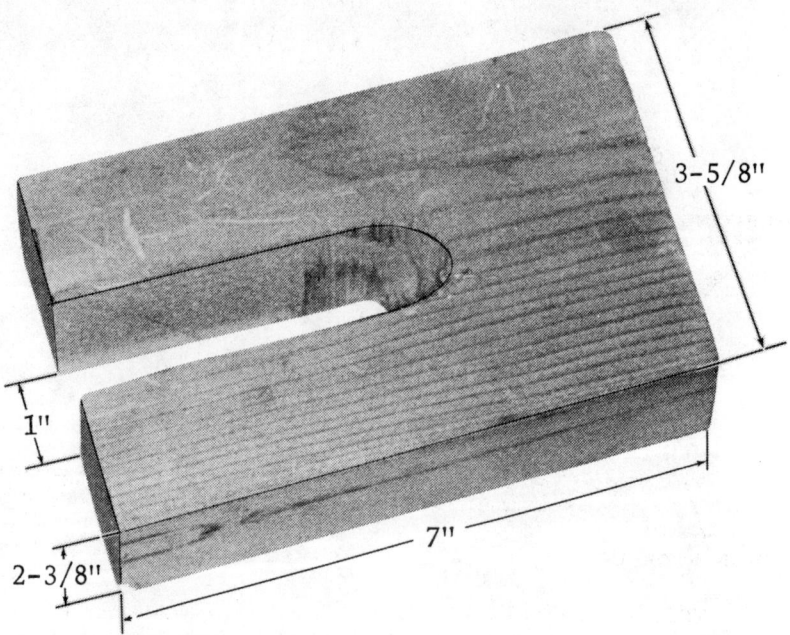

These are the dimensions of the Cruise Control (wooden block) that the author made for less than $1.00. Your block may have to have slightly different dimensions.

CRUISE CONTROL

This automatic speed-control accessory costs about $95.00 and can be a worthwhile investment as an add-on device, provided your driving is mostly at higher speeds and especially if you have a "heavy foot" on the accelerator pedal. This accessory maintains any speed you set it for without your touching the accelerator pedal, and this is where the fuel savings come by minimizing throttle movement.

A recently installed Cruise Control saved a Chevrolet owner 5 mpg on a 500-mile trip that he made at 55 mph. This driver had a "heavy foot" and he never got over 15 mpg before he installed this device. The 1/3 improvement saved him $10.00 on this one trip alone. If you frequently drive on long trips, then the savings can be

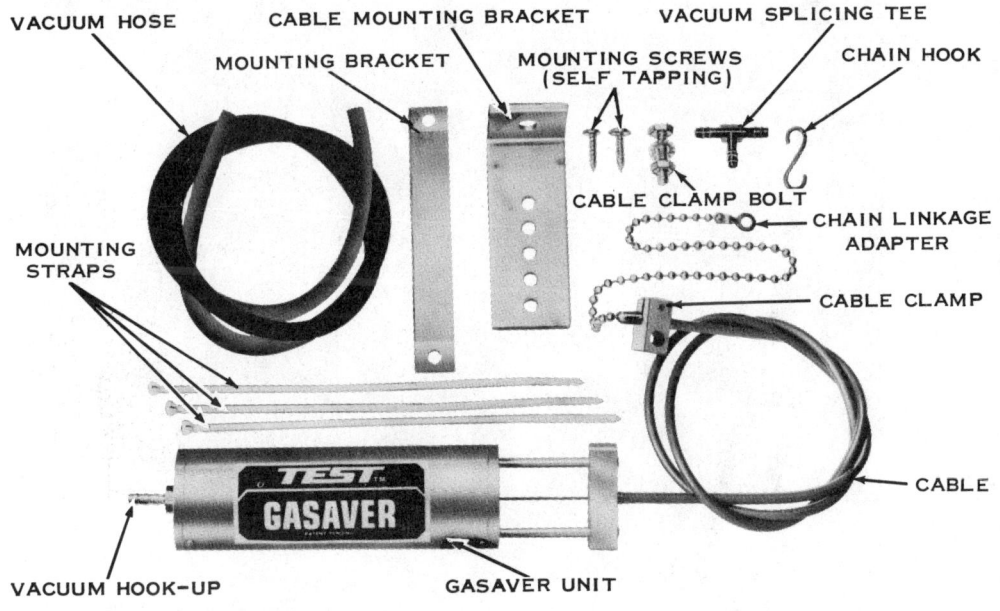

The TEST gasaver kit sells for less than $30.00. It acts like a cruise control in maintaining a constant speed. It offers resistance to the pressure you exert on the accelerator pedal, helping you to avoid jack-rabbit starts and sudden speed spurts. The unit is available from Tanner Electronic Systems Inc., 19428 Londelius St. Northridge CA. 91324.

The TEST gasaver installed on the tire well of the engine compartment offers resistance to accelerator movement that can be adjusted to the level desired. The kit comes with an instruction booklet so that any amateur can install it quickly.

considerable and will pay for the device in a short time. If your driving is mostly in the city, then this device is worthless, because you must constantly change your speed to keep up with the flow of traffic.

AIR CONDITIONER

If you have a car with an air conditioner, you're dragging around about 100 pounds of extra weight that's costing you about 0.2 mpg (at 77¢ per month) even when you don't use it. And when you do use it, you're losing another 2.5 mpg for comfort. That's a monthly penalty of about 16 gallons of fuel at $10.00.

It is worth it? You'll have to be the judge. However, you can use it judiciously and only when its so hot outside that opening the windows and air vents do not provide the necessary comfort.

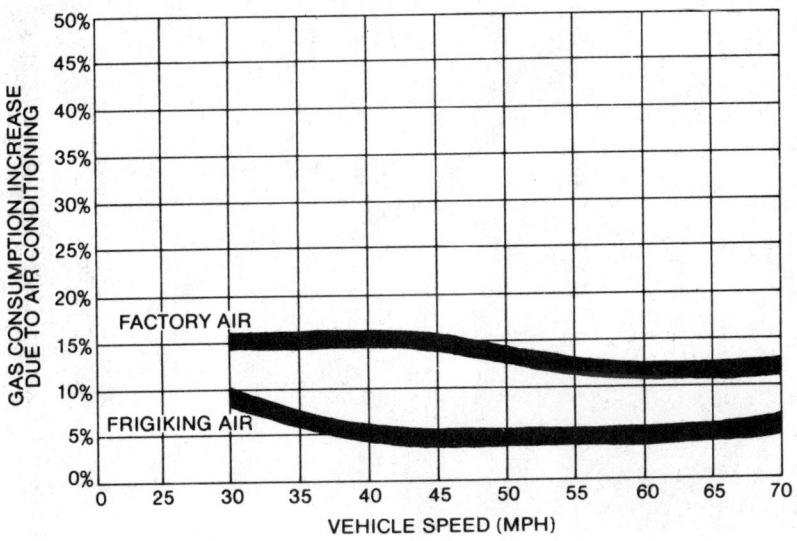

Some interesting tests have been made as to the gas consumption increase due to the use of an air conditioner. This graph shows the different effects of two basic types of systems, one in which the compressor is working continuously (Factory Air), and one in which the compressor clutch cycles and works only when cooling is demanded (Frigiking Air). These graphs represent the average figures obtained from testing six different cars.

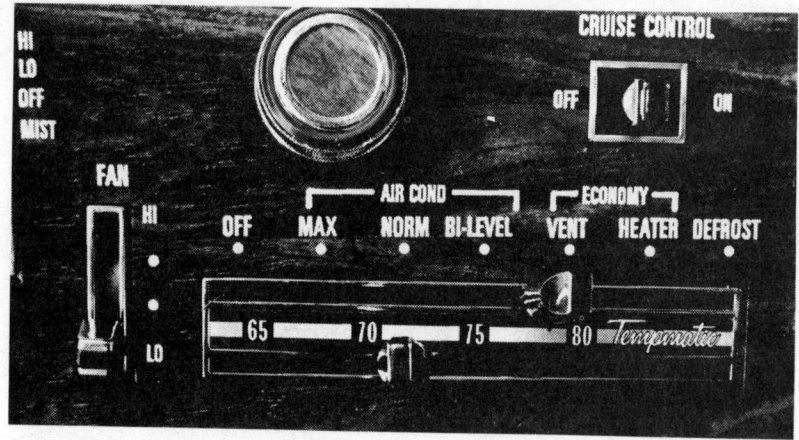

The 1975 Oldsmobile has an "Economy" range setting for the air conditioner. This setting opens vents to increase the air flow when the compressor is not needed.

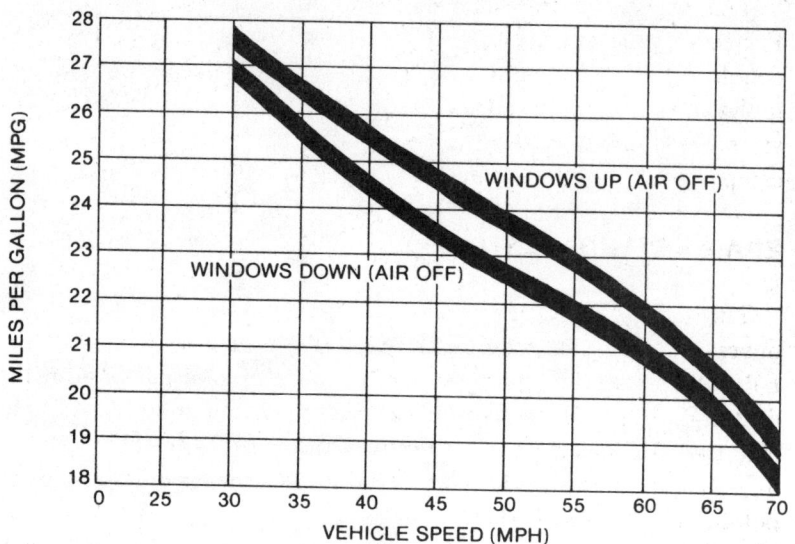

Window position has a significant effect on gas consumption because it changes the aerodynamic drag. In a test of six different vehicles with the windows up and down, Frigiking drew these graphs which showed that the gas consumption increased an average of 2.8% with the windows down. One car used 7.5% more gas with its windows down than it did with them up.

appendix

This section of the book contains some EPA (Environmental Protection Agency) gas mileage predictions for 1975 model cars and details of their labeling program. In addition, three very important tables are supplied for determining your costs and savings under various improvement factors. These tables will be handy for making comparisons.

EPA LABELING PROGRAM

The EPA has developed a special program to help a new car buyer select an automobile with good fuel economy. Basically, the EPA was created to clean the air in our cities, and one of its main directions has been to reduce emissions from automobiles. To this end, the EPA has been in a running battle with Detroit to force the manufacturers reduce the size of engines as a means of lessening air pollution. Since 1974, they have "asked" auto manufacturers to display one of two labels indicating the kind of gas mileage that a car buyer can expect: The Specific Label and The Comparative Label.

The test procedures used to provide the information on the labels involves a 23-minute, 7.5-mile test under simulated commuter-type driving conditions. Top speed in the test is 57 mph, and the average

150 GAS GUZZLER'S GUIDE

is about 20 mph. A dynamometer is used to insure results that are scientifically comparable. However, if the cars were actually driven at a constant speed on the highway, their miles-per-gallon rating would improve, but the comparison between the weight categories would remain approximately the same.

THE SPECIFIC LABEL

This label gives the miles-per-gallon performance for the specific vehicle on which it is displayed. The label also gives comparative information about weight classes, range of miles per gallon, average miles per gallon, and fuel costs. As a prospective buyer, you will be able to see how the fuel economy of a particular automobile compares with cars in other weight classes.

Vehicle Test Weight (lbs)	Range of MPG	Average MPG	Fuel Costs 10,000 mi. and 40¢/gal.
2,000	22-29	24	$165
2,250	19-25	21.5	$185
2,500	17-22.5	18.5	$215
2,750	10.5-24.5	17.5	$230
3,000 [3,100]	9-20	15 [17]	$265 [$235]
3,500	10.5-20	13.5	$295
4,000	6.5-19	10.5	$380
4,500	7.5-14	9.5	$420
5,000	7-11	9	$445
5,500	7-10.5	8	$500

This EPA label illustrates the type of information which would be presented on each car. This particular vehicle weighs 3,100 pounds and registered an average fuel economy figure of 17 mpg in the EPA test. Note that this figure is 2 mpg better than the average for the car's weight class. The buyer can expect to save about $90.00 yearly in gasoline costs over the average vehicle in its weight class.

The Comparative Label

This label includes the same fuel-economy table as the previous one. The measured miles-per-gallon performance of the individual car, however, is not indicated. Instead, the weight class to which the car belongs is circled. This label will allow you to make general comparisons between weight classes and the corresponding fuel economy.

1975 GAS MILEAGE GUIDE FOR NEW CAR BUYERS

The EPA in cooperation with the Federal Energy Administration has prepared the following tables for you to compare miles-per-gallon information for the broad range of automobiles expected to be sold in this country in 1975. The cars used were prototypes which the EPA tested in its own laboratory to assure compliance with air pollution standards. Because the same engines are used in a number of different cars, it is not necessary to test each particular model. The fuel economy figures are estimates, therefore, this does not mean that you will necessarily get the same mileage. Many factors affect the mileage, and a great number of them are under the direct control of the driver as repeatedly pointed out in this book.

Most cars have a label in the rear window to indicate the fuel economy of that vehicle. In some cases, this figure will not agree with those in the following list. This is because some manufacturers have elected to give more detailed information that is specific to the weight, transmission, and rear axle ratio of the individual car, as well as to the car line, engine size, fuel system, and catalyst usage. Fuel economy figures based on this type of detailed car description **are more precise** than those listed here since more factors about the car are taken into consideration. California cars are often different from those sold in the 49 states.

If you wish to obtain these figures, write a letter requesting California car information to: Fuel Economy, Pueblo, Colorado 81009.

GAS GUZZLER'S GUIDE

1975

Manufacturer/Car line	Engine size (cu. in. disp.)	Cylinders	Carburetor (barrels/fuel inj.)	Catalyst	Fuel economy (miles per gal.) City	Hwy.
American Motors						
Gremlin	232	6	1	X	17	23
	258	6	1	X	14	20
	304	8	2	X	13	21
Hornet	232	6	1	X	15	19
	258	6	1	X	14	22
	304	8	2	X	13	21
Hornet Wagon	232	6	1	X	15	19
	258	6	1	X	14	22
	304	8	2	X	13	21
Matador	304	8	2	X	13	18
	360	8	2	X	12	16
	360	8	4	X	11	16
	401	8	4	X	11	14
Matador Wagon	304	8	2	X	13	18
	360	8	2	X	12	16
	360	8	4	X	11	16
	401	8	4	X	11	14
Audi						
Fox	97	4	FI	X	22	33
100	114	4	FI	X	18	27
BMW						
2002	121	4	2		19	30
530	182	6	FI		12	15
3.0S	182	6	FI		12	15
Buick						
Apollo	250	6	1	X	15	20
Skylark	231	6	2	X	15	21
Apollo/Skylark	260	8	2	X	13	17
	350	8	4	X	14	18
Skyhawk	231	6	2	X	17	23
Century/Regal	231	6	2	X	14	19
	350	8	4	X	13	20
Century Wagon	350	8	4	X	12	16
Lesabre	350	8	4	X	12	16
	455	8	4	X	11	16
Estate Wagon	455	8	4	X	10	14
Electra	455	8	4	X	10	13
Riviera	455	8	4	X	10	13
Cadillac						
Cadillac	500	8	4	X	11	14
Fleetwood 75	500	8	4	X	10	13
Eldorado	500	8	4	X	11	14

1975 GAS MILEAGE GUIDE

Manufacturer/Car line	Engine size (cu. in. disp.)	Cylinders	Carburetor (barrels/fuel inj.)	Catalyst	Fuel economy (miles per gal.) City	Hwy.
Chevrolet						
Vega	140	4	2	X	19	28
Vega Kammback	140	4	2	X	18	28
Monza	140	4	2	X	18	28
Nova	250	6	1	X	15	20
	350	8	4	X	13	18
Camaro	250	6	1	X	15	20
	350	8	4	X	13	18
Chevelle	350	8	4	X	12	16
	400	8	4	X	11	16
Malibu Wagon	350	8	4	X	11	16
	400	8	4	X	11	14
Chevrolet	350	8	4	X	11	16
	400	8	4	X	11	14
Chevrolet Wagon	400	8	4	X	10	13
Monte Carlo	350	8	4	X	12	16
	400	8	4	X	11	16
Corvette	350	8	4	X	13	18
Chrysler						
Cordoba	318	8	2	X	11	14
	360	8	4	X	11	19
	400	8	4	X	12	17
Chrysler	360	8	4	X	11	15
	400	8	4	X	10	15
	440	8	4	X	9	13
Chrysler Wagon	400	8	4	X	9	13
	440	8	4	X	9	13
Imperial	440	8	4	X	9	13
Datsun						
B-210	85	4	2	X	27	39
710	119	4	2	X	21	29
710 Wagon	119	4	2	X	18	24
610	119	4	2	X	18	24
610 Wagon	119	4	2	X	18	24
Dodge						
Dart	225	6	1	X	15	19
	318	8	2	X	12	16
	360	8	4	X	11	17
Coronet/Charger	318	8	2	X	11	14
	360	8	4	X	11	19
	400	8	4	X	12	17

1975

Manufacturer/Car line	Engine size (cu. in. disp.)	Cylinders	Carburetor (barrels/fuel inj.)	Catalyst	Fuel economy (miles per gal.) City	Hwy.
Coronet Wagon	360	8	4	X	11	15
	400	8	4	X	10	15
Monaco	360	8	4	X	11	15
	400	8	4	X	10	15
	440	8	4	X	9	14
Monaco Wagon	360	8	4	X	10	15
	400	8	4	X	9	13
	440	8	4	X	9	13
Ford						
Pinto	140 (2.3L)	4	2	X	17	25
	171 (2.8L)	6	2	X	14	21
Pinto Wagon	140 (2.3L)	4	2	X	17	25
	171 (2.8L)	6	2	X	14	21
Mustang II	140 (2.3L)	4	2	X	17	25
	171 (2.8L)	6	2	X	14	21
	302	8	2	X	12	16
Maverick	250	6	1	X	15	21
	302	8	2	X	12	16
Granada	250	6	1	X	15	20
	302	8	2	X	10	14
	351	8	2	X	9	13
Torino/Elite	351	8	2	X	10	16
	400	8	2	X	9	14
	460	8	4	X	11	13
Torino Wagon	351	8	2	X	10	16
	400	8	2	X	9	14
	460	8	4	X	9	13
Ford	351	8	2	X	10	16
	400	8	2	X	9	14
	460	8	4	X	10	13
Ford Wagon	400	8	2	X	9	14
	460	8	4	X	9	13
Thunderbird	460	8	4	X	9	13
Lincoln-Mercury						
Comet	250	6	1	X	15	21
	302	8	2	X	12	16
Monarch	250	6	1	X	15	20
	302	8	2	X	10	14
	351	8	2	X	9	13
Montego/Cougar	351	8	2	X	10	16
	400	8	2	X	9	14
	460	8	4	X	11	13

1975 GAS MILEAGE GUIDE

Manufacturer/Car line	Engine size (cu. in. disp.)	Cylinders	Carburetor (barrels/fuel inj.)	Catalyst	Fuel economy (miles per gal.) City	Hwy.
Montego Wagon	351	8	2	X	10	16
	400	8	2	X	9	14
	460	8	4	X	9	13
Mercury	400	8	2	X	9	14
	460	8	4	X	9	13
Mercury Wagon	400	8	2	X	9	14
	460	8	4	X	9	13
Lincoln, Continental	460	8	4	X	9	13
Continental Mark IV	460	8	4	X	9	13
Mercedes-Benz						
240D	147	4	FI		24	31
300D	183	5	FI		24	31
230	141	4	1	X	16	20
280/280C	167	6	4	X	15	20
280S	167	6	4	X	15	20
450 SE/SEL	276	8	FI	X	11	17
450 SL/SLC	276	8	FI	X	11	17
Oldsmobile						
Omega	250	6	1	X	15	20
	260	8	2	X	13	17
	350	8	4	X	14	18
Starfire	231	6	2	X	17	23
Cutlass	260	8	2	X	13	17
	350	8	4	X	13	19
	455	8	4	X	11	19
Cutlass Wagon	350	8	4	X	12	16
	455	8	4	X	11	16
Delta 88	350	8	4	X	12	16
	455	8	4	X	11	16
Custom Cruiser Wagon	455	8	4	X	11	16
Olds 98	455	8	4	X	10	15
Toronado	455	8	4	X	10	15
Peugeot						
504	120	4	2		19	26
504 Wagon	120	4	2		17	22
Plymouth						
Valiant/Duster	225	6	1	X	15	20
	318	8	2	X	12	16
	360	8	4	X	11	17

GAS GUZZLER'S GUIDE

1975

Manufacturer/Car line	Engine size (cu. in. disp.)	Cylinders	Carburetor (barrels/fuel inj.)	Catalyst	Fuel economy (miles per gal.) City	Fuel economy (miles per gal.) Hwy.
Road Runner, Fury	318	8	2	X	11	14
	360	8	4	X	11	19
	400	8	4	X	12	17
Fury Wagon	360	8	4	X	11	15
	400	8	4	X	10	15
Gran Fury	360	8	4	X	11	15
	400	8	4	X	10	15
	440	8	4	X	9	14
Gran Fury Wagon	360	8	4	X	10	15
	400	8	4	X	9	13
	440	8	4	X	9	13
Pontiac						
Astre	140	4	2	X	18	28
Astre Wagon	140	4	2	X	18	28
Ventura	250	6	1	X	15	20
	260	8	2	X	13	17
	350	8	4	X	14	18
Firebird	250	6	1	X	15	20
	350	8	4	X	11	15
	400	8	4	X	11	15
Lemans	350	8	4	X	10	13
Lemans/Grand AM	400	8	4	X	11	15
Grand AM	455	8	4	X	10	14
Lemans Wagon	400	8	4	X	11	15
Pontiac	400	8	4	X	11	15
	455	8	4	X	10	14
Pontiac Wagon	455	8	4	X	9	12
Grand Prix	400	8	4	X	11	15
	455	8	4	X	10	14
Porsche						
914	109 (1.8L)	4	FI	X	19	32
	120 (2.0L)	4	FI	X	20	28
Saab						
99	121	4	FI		21	25
Toyota						
Corolla	97	4	2	X	20	28
Corolla Wagon	97	4	2	X	20	28
Corona	133	4	2	X	18	25
Corona Wagon	133	4	2	X	18	25
Celica	133	4	2	X	17	27
Corona Mk. II	156	6	2	X	16	19
Corona Mk. II Wagon	156	6	2	X	16	19
Volkswagen						
Beetle	97	4	FI	X	23	34
Rabbit	90	4	2	X	24	38
Dasher	90	4	2	X	23	35
Dasher Wagon	90	4	2	X	23	35
Scirocco	90	4	2	X	24	38
Thing	97	4	FI	X	23	34
Volvo						
240	121	4	FI	X	16	26
245 Wagon	121	4	FI	X	16	25
160	182	6	FI	X	16	26

SPARK PLUG TORQUE SPECIFICATIONS 157

GASKET TYPE

SPARK PLUG SIZE	CAST IRON HEAD		ALUMINUM HEAD	
	TORQUE (FT-LBS)	FINGER TIGHT, PLUS (TURNS)	TORQUE (FT-LBS)	FINGER TIGHT, PLUS (TURNS)
10MM	8-12	1/4	8-12	1/4
12MM	10-18	1/4	10-18	1/4
14MM	26-30	1/4	18-22	1/4
18MM	32-38	1/4	28-34	1/4

TAPER SEAT TYPE

SPARK PLUG SIZE	CAST IRON HEAD		ALUMINUM HEAD	
	TORQUE (FT-LBS)	FINGER TIGHT, PLUS (TURNS)	TORQUE (FT-LBS)	FINGER TIGHT, PLUS (TURNS)
10MM	-	-	-	-
12MM	-	-	-	-
14MM	10-20	1/16	10-20	1/16
18MM	15-20	1/16	15-20	1/16

IF YOU IMPROVE YOUR GAS MILEAGE, YOUR SAVINGS WILL BE:

AT 10% IMPROVEMENT GALLONS SAVED	DOLLARS SAVED	AT 15% IMPROVEMENT GALLONS SAVED	DOLLARS SAVED	AT 20% IMPROVEMENT GALLONS SAVED	DOLLARS SAVED
Yearly:		**Yearly:**		**Yearly:**	
100	60.00	150	90.00	200	120.00
Monthly:		**Monthly:**		**Monthly:**	
8.33	5.00	12.5	7.50	16.66	10.00
Daily:		**Daily:**		**Daily:**	
0.28	0.17	0.42	0.25	0.56	0.34

IF YOU IMPROVE YOUR GAS MILEAGE, YOUR SAVINGS WILL BE:

AT 25% IMPROVEMENT GALLONS SAVED	DOLLARS SAVED	AT 30% IMPROVEMENT GALLONS SAVED	DOLLARS SAVED	AT 35% IMPROVEMENT GALLONS SAVED	DOLLARS SAVED
Yearly:		**Yearly:**		**Yearly:**	
250	150.00	300	180.00	350	210.00
Monthly:		**Monthly:**		**Monthly:**	
20.8	12.50	25	15.00	29.2	17.50
Daily:		**Daily:**		**Daily:**	
0.70	0.42	0.83	0.50	0.97	0.48

MILEAGE IMPROVEMENT DATA

IF YOU IMPROVE YOUR GAS MILEAGE, YOUR SAVINGS WILL BE:

AT 40% IMPROVEMENT		AT 45% IMPROVEMENT		AT 50% IMPROVEMENT	
GALLONS SAVED	DOLLARS SAVED	GALLONS SAVED	DOLLARS SAVED	GALLONS SAVED	DOLLARS SAVED
Yearly: 400	240.00	**Yearly:** 450	270.00	**Yearly:** 500	300.00
Monthly: 33.3	20.00	**Monthly:** 37.5	22.50	**Monthly:** 41.7	25.00
Daily: 1.1	0.67	**Daily:** 1.25	0.75	**Daily:** 1.4	0.83

IF YOU IMPROVE YOUR GAS MILEAGE, YOUR SAVINGS WILL BE:

AT 1 MPG IMPROVEMENT		AT 2 MPG IMPROVEMENT		AT 3 MPG IMPROVEMENT	
GALLONS SAVED	DOLLARS SAVED	GALLONS SAVED	DOLLARS SAVED	GALLONS SAVED	DOLLARS SAVED
Yearly: 77	46.20	**Yearly:** 144	92.40	**Yearly:** 231	138.60
Monthly: 6.4	3.84	**Monthly:** 12.8	7.68	**Monthly:** 19.25	11.55
Daily: 0.21	0.13	**Daily:** 0.42	0.26	**Daily:** 0.64	0.39

IF YOU IMPROVE YOUR GAS MILEAGE, YOUR SAVINGS WILL BE:

AT 4 MPG IMPROVEMENT		AT 5 MPG IMPROVEMENT		AT 6 MPG IMPROVEMENT	
GALLONS SAVED	DOLLARS SAVED	GALLONS SAVED	DOLLARS SAVED	GALLONS SAVED	DOLLARS SAVED
Yearly:		**Yearly:**		**Yearly:**	
288	172.80	385	231.00	462	277.20
Monthly:		**Monthly:**		**Monthly:**	
25.6	14.40	32.0	19.25	38.5	23.10
Daily:		**Daily:**		**Daily:**	
0.84	0.48	1.07	0.64	1.28	0.77

AVERAGE FUEL CONSUMPTION BASED ON MILES PER GALLON

AT 12 MILES PER GALLON		AT 13 MILES PER GALLON		AT 14 MILES PER GALLON	
GALLONS	COST	GALLONS	COST	GALLONS	COST
Yearly:		**Yearly:**		**Yearly:**	
1000	600.00	923	553.84	857	514.29
Monthly:		**Monthly:**		**Monthly:**	
83.3	50.00	76.9	46.20	71.4	42.86
Daily:		**Daily:**		**Daily:**	
2.78	1.67	2.56	1.54	2.38	1.43

MILEAGE IMPROVEMENT DATA

AVERAGE FUEL CONSUMPTION BASED ON MILES PER GALLON

AT 15 MILES PER GALLON		AT 16 MILES PER GALLON		AT 17 MILES PER GALLON	
GALLONS	COST	GALLONS	COST	GALLONS	COST
Yearly:		**Yearly:**		**Yearly:**	
800	480.00	750	450.00	706	423.53
Monthly:		**Monthly:**		**Monthly:**	
66.7	40.00	62.5	37.50	58.8	35.29
Daily:		**Daily:**		**Daily:**	
2.22	1.33	2.08	1.25	1.96	1.18

AVERAGE FUEL CONSUMPTION BASED ON MILES PER GALLON

AT 18 MILES PER GALLON		AT 19 MILES PER GALLON		AT 20 MILES PER GALLON	
GALLONS	COST	GALLONS	COST	GALLONS	COST
Yearly:		**Yearly:**		**Yearly:**	
667	400.00	632	378.95	600	360.00
Monthly:		**Monthly:**		**Monthly:**	
55.6	33.33	52.6	31.58	50	30.00
Daily:		**Daily:**		**Daily:**	
1.85	1.11	1.75	1.05	1.67	1.00

index

A

Accelerating system   73
Accelerator pump plunger   73
Accessories that affect gas mileage   139
Adjusting the automatic choke   76
Adjustment, valve   16
Adjusting the carburetor, using the lean-drop method   79
Adjusting the float level   69
Advance, centrifugal   33, 34
Advance mechanisms, automatic   33, 34
Advance, vacuum   36
Aerodynamic drag   103, 107, 148
Air bleed   71
Air cleaner, replace   49
Air conditioner   147
Air cleaner, thermostatically controlled   57
Air filter   52
Air filter, cleaning   55, 56
Air filter element, clogged   54
Air-fuel mixture   19, 58, 61, 62
Air-fuel mixture, rich   69
Air-fuel mixture, testing   64
Air inflation, tires   109, 111
Aligning contact points   30
Allignment, wheel   116, 117
Alternator   16, 128

Arabs   1, 2
Automatic advance mechanisms   33, 34
Automatic choke   62, 74
Automatic choke, sticking   75, 76
Average fuel consumption costs   3

B

Base mileage   9
Basic mileage test   8, 12
Basic tests, spark-delay system   98
Basic tests, transmission-controlled spark system   89, 91
Basic tools   12
Battery   16
Battery cable, corroded   18
Battery terminals   17
Battery terminals, corrosion   19
Battery, testing   18
Belted bias tire   112
Bias ply tires   112
Bicycle racks   108
Boot, spark plug   24
Brakes, disc   117
Brakes, drum-type   117
BTU   4
Burned valves   61
Buying gasoline wisely   118
Buying oil wisely   122
Buying spark plugs   22

INDEX

C

Cam angle 32
Carburetors 62
Carburetor adjustments 77
Carburetor bench adjustments 65
Carburetor circuits 69, 70
Carburetor flow curve 63
Carburetor kits 65
Carburetor, operation of 63
Carburetor, service procedures 64
Carburetor specifications 65
Catalytic converter 119, 121
Centrifugal advance 33, 34
Centrifugal advance mechanism, testing 45
Changing the pump stroke 74
Cheap energy 1
Chrysler Fuel-Pacer system 141
Clogged air filter element 54
CO emissions 53
Cold-air kit 60
CO meter 81
Contact points 17
Contact points, adjusting the gap 31
Contact points, aligning 30
Contact points, defective 29
Contact points, replacing 28
Contact points, spacing 32
Cooling system 127
Combustion chamber 19
Combustion temperature 19
Compression, testing 24, 25
Crankcase emissions 81
Crankcase emission-control systems 82
Crankcase emission-control system, quick test of 82
Crankcase emission-control system, service of 84
Cruise control 144, 145
Cruise control for $1.00 144
Cylinder head nuts, tighten 15

D

Defective air filter, results of 56

Defective automatic choke 75
Defective distributor cap 41
Defective fuel pump diaphragm 52
Defective gasket 72
Defective high-tension wires 42
Defective idle mixture adjusting needle 68
Detonation 61, 119, 120, 121
Diaphragm 72
Diaphragm, worn 66
Disc brakes 117
Distributor 16
Distributor cap, cracked 41
Distributor cap, servicing 40
Distributor, replacing 41
Distributor, servicing 26
Distributor, V-8 engine 28
Drag efficiency 107
Driving in hilly country 136
Driving skills 124
Driving skills, improving 125
Driving the freeways 135
Drum-type brakes 117
Dwell 32
Dwell, adjusting 33

E

Economy run 134
EGR system quick tests 101
Electrolyte 18
Emission-controlled engines 41, 61, 79
Emission-controlled engines, tuning 81
Emission-control system 12, 13, 81
Emission-control system, evaporative 80
Emissions, HC & CO 53
Energy, gasoline, heat, mechanical 5
Engine, cold, starting 126
Engine, cooling system 127
Engine, internal-combustion 10
Engine, performance-effectiveness 61
Engine, power-timing 45
Engine, steam 10
Engine, tune-up 13

164 INDEX

Engines, emission-controlled 47, 79
Environmental Protection Agency 1
EPA 149
EPA labeling program 149
  Specific label 150
  Comparative label 151
Evaporative emissions 81
Evaporative emission-control system 80
Excessive idling 130
Exhaust emissions 80, 81
Exhaust emission-control systems 86
Exhaust-gas recirculation 100
Exhaust manifold 14, 15
Extra weight 105

F

Fan belts, worn 13
Filter element, crankcase emission-control system 85
Filter element, replace 69
Flash-over, spark plug 26
Flat spot during acceleration 73
Float level, adjusting 69
Float system 69
Footprint, tire 113, 115
Freeways, driving 135
Fuel filter, replacing 49, 53, 54
Fuel pump diaphragm, defective 52
Fuel consumption costs 3
Fuel system 50
Fuel system service 13, 49
Fuel, wasting 20

G

Gasket, defective 13
Gasolines 118
Gasoline, buying wisely 118
Gas-consumption variables 3
Gas mileage, air cleaner 49
Gas mileage, calculation of 9
Gas mileage guide (1975) 151
Gas mileage improvement schedules 158
Gas mileage improvers 59
Gas mileage, measurement of 6
Gas mileage, vehicle's potential 9

Gas tank, filling 7
Getting underway 131
Gravity 146, 136

H

HC emissions 53
Heat-control valve 15
Heat range, spark plug 23
Hilly country, driving in 12, 136
High-tension wires, defective 42
High-test gasoline 118
Horsepower, maximum 19
Hot-rod starts 1, 131
Honda CVCC engine 80
Hydraulic valve lifters 16
Hydrocarbon emissions 47, 81

I

Idling, excessive 130
Idle system 71
Idle mixture adjusting needle, defective 68
Ignition coil 16, 42
Ignition cross-firing 44, 45
Ignition repairs 13
Ignition system 16, 17
Ignition timing 17
Ignition timing, adjusting 43
Improving your driving skills 125
Inflation decal 110, 111
Insulator, spark plug 19
Intake manifold bolts, tighten 15
Intake manifold gasket, defective 13
Internal-combustion engine 10, 11

J

Jack-rabbit starts 1, 125

K

Kinetic energy 14, 137, 138

L

Leaded gasoline 119
Lean air-fuel mixture 65
Lean-drop method of adjusting the carburetor 79, 81

INDEX 165

Liquid fuel 15
Low-lead gasoline 118
Lubricant, Lube Trap 15

M

Main metering system 72
Manifold, exhaust 14
Manifold heat control valve 14
Mechanical conditions 13, 15
Mechanical tappets 16
Metering jet 72
Metering rod 66, 72
Micky Thompson 134
Mid-eastern nations 1
Mileage, base 9
Miles per gallon, 164.5 4
Mileage test, basic 8
Misfiring spark plugs 19, 20, 21, 22 25

N

Needle valve, worn 67
Needle valve and seat assembly 69
Normal spark plug 25
NOx exhaust emissions 80, 100
NOx system 86

O

Octane number 118, 119, 121
Odometer 9
Oil, buying wisely 122
Oil viscosity 122
Old Smoothie 133
Orifice spark-control system 97
Overheated spark plug 25

P

Pacing traffic lights 132
PCV valve 82
Pinging 120, 121
Piston ring trouble 26
Ported spark advance 37
Power jet 72
Power system 72
Power-timing the engine 45
Preignition 23, 46, 47, 60
Premium gasoline 118, 119

Preplanning 130
Pump stroke, changing 74

Q

Quick tests, EGR system 101
Quick test of the crankcase emission-control system 82

R

Rationing 1
Radial ply tires 112
Reducing hydrocarbon emissions 47
Reducing tire rolling resistance 114
Regular gasoline 120
Removing spark plugs 23
Replacing the contact points 28
Replacing the distributor 41
Replacing the fuel filter 54
Replacing the spark plugs 22, 43
Retrofit system 98
Rich air-fuel mixture 65
Road surfaces 117
Rolling resistance 103, 109, 135
Roof racks 108

S

Sea-sawing 124
Servicing the automatic advance mechanisms 40
Servicing the carburetor 64
Servicing the crankcase emission-control system 84
Servicing the distributor 26
Servicing the distributor cap 40
Servicing the fuel system 49
Setting the ignition timing 43
Servicing the spark-delay system 99
Servicing the thermostatically-controlled air cleaner 59
Snow tires 115
Spark advance, ported 37
Spark-delay valve 96
Spark delay systems 97
Spark plugs 16, 17, 20
Spark plug boot 24
Spark plug, filing the electrodes 27

166 INDEX

Spark plug flash-over 26
Spark plug heat range 23
Spark plug insulator 19
Spark plug misfiring 19, 21, 22, 25
Spark plug, normal 25
Spark plug, overheated 25
Spark plug, pre-ignition 48
Spark plugs, removing 23
Spark plugs, replacing 43
Spark plugs, selecting 22
Spark plug, setting the gap 26
Spark plug torque specifications 157
Spark plug, types 22
Spark plug, worn 27
Speed-controlled spark system 86
Starting a cold engine 126
Steam engine 10
Stop-and-go driving 124
System isolation test, Chrysler's NOx system 95
System isolation test, transmission-controlled spark system 93

T

Tappets, machanical 16
Temperature, combustion 19
TEST Gasaver 145
Testing the air fuel mixture 64
Testing the centrifugal advance mechanism 40, 45
Testing the compression 24
Testing the vacuum advance mechanism 40
Tetraethyl lead 121
Thermal-Vacuum-Switching valve (TVS) 38, 39
Thermostat 128, 130
Thermostatically-controlled air cleaner 57, 59
Throttle shaft, worn 65
Time-delay relay 90, 91
Timing indicator 43
Timing, ignition 17
Timing light 12, 46
Tires 109
Tire air inflation 109, 111
Tire-screeching stops 1

Tools, basic 12
Top Dead Center (TDC) 33
Transmission-controlled spark system 86
Transmission-controlled spark system isolation test 93
Transmission-controlled spark system, troubleshooting 92
Transmission-regulated spark system 86
Troubleshooting the transmission-controlled spark system 92
Tuning the engine 10, 13

U

Unfiltered air, 56, 57
Unleaded gasoline 118

V

Vacuum advance 36, 39
Vacuum gauge 12, 45, 48, 77, 81, 89, 131, 139
Vacuum-Switching valve 38, 39
Valve adjustment 16
Valve, heat-control 15
Valve lash 16
Valve, spark delay 96
Valve, thermostatically controlled 15
Valve timing 16
Valve trouble 26
Vehicle weight 103, 104
Vinyl top 108
Viscosity, oil 122
Volumetric efficiency 61, 121

W

Wasting fuel 20
Weight 103
Wheel alignment 116, 117
Wind resistance 103, 106, 135
Wooden block 142
Worn fan belt 13
Worn needle valve 67
Worn spark plug 27

Z

Zip kit 65